access to geography

POPULATION

304.6

Jack Gillett

Hodder Murray

A MEMBER OF THE HODDER HEADLINE GROUP

For Roger and Philippa Small, Lytham St Annes

The Publishers would like to thank the following for permission to reproduce copyright illustrations: p.53 © Corbis/Luc Gnago/Reuters; **p.61** © Getty Images/ Vano Shlamov; **p.78** © Alamy/David Lyons; **p.138** © Robert Harding World Imagery/ Marco Simoni.
The Publishers would also like to thank the following for permission to reproduce copyright material: p.27 © *The Guardian* for an extract from *Shanghai eases China's one-child rule* by Jonathan Watts (14 April 2004); **pp.95–6** © Telegraph Group Limited for an extract from *East Germany's vision of unity fades* by Kate Connolly (11 September 2004); **pp.51–2** © Telegraph Group Limited for an extract from *The children of rape* by Philip Sherwell (25 July 2004).
Every effort has been made to trace all copyright holders, but if any have been inadvertently overlooked the Publishers will be pleased to make the necessary arrangements at the first opportunity.

Although every effort has been made to ensure that website addresses are correct at time of going to press, Hodder Murray cannot be held responsible for the content of any website mentioned in this book. It is sometimes possible to find a relocated web page by typing in the address of the home page for a website in the URL window of your browser.

Orders: please contact Bookpoint Ltd, 130 Milton Park, Abingdon, Oxon OX14 4SB. Telephone: (44) 01235 827720. Fax: (44) 01235 400454. Lines are open 9.00–6.00, Monday to Saturday, with a 24-hour message answering service. Visit our website at www.hoddereducation.co.uk

© Jack Gillett 2005
First published in 2005 by
Hodder Murray, an imprint of Hodder Education,
a member of the Hodder Headline Group
338 Euston Road
London NW1 3BH

Impression number	10 9 8 7 6 5 4 3 2 1
Year	2010 2009 2008 2007 2006 2005

Cover photo © Mark Edwards/Still Pictures
Typeset in 10/11pt Baskerville and produced by Gray Publishing, Tunbridge Wells
Printed in Malta

A catalogue record for this title is available from the British Library

ISBN-10: 0340 886 730
ISBN-13: 978 0340 886 731

Contents

Introduction

This book is about demography, a branch of social sciences that is concerned with population characteristics and trends. It has been an increasingly important geographical theme since the 1950s and is now a major component of all the AS/A2 specifications for this subject, reflecting the wide range of global population issues such as those listed below. The importance of these issues is also reflected in the time and financial investment that non-governmental organisations such as Oxfam, individual countries and global-scale institutions such as the World Bank devote to them.

- The rapid increase in global population, particularly during the 'population explosion' of the twentieth century.
- The global population now exceeds six billion and is predicted to continue rising at a rate which might prove unsustainable in terms of food and other essentials.
- The occurrence of pandemics such as the current HIV/AIDS crisis.
- The challenge of increasing longevity and meeting the needs of ageing populations.
- The accelerating rate of population migration. Recent, and possibly conservative, estimates of the number of annual global migrants exceed 150 million.
- The increasing disparity of the quality of life between the more and less fortunate people within individual countries as well as on a global scale.
- The cycles of natural and human-induced disasters that continue to afflict particular regions of the world, e.g. flooding in Bangladesh and famine in many African regions.

An even more important function of this book is to stimulate a life-long interest all matters relating to the human condition such as population location, mobility, economic status and quality of life. The importance of these topics is constantly reinforced by the media coverage devoted to them; newspaper headlines frequently highlight population-related issues and some of the television's most memorable images have been based on them. The media coverage of the 1984 famine in Ethiopia that resulted in Sir Bob Geldof's 'Live Aid' fundraising concerts remains a classic example of media potential to highlight problems and initiate appropriate remedial measures.

Each of the book's chapters deals with a key population-related issue, even though the themes of the chapters are very closely linked. For example, migration patterns are often influenced by human conflict and variations in quality of life. Each chapter incorporates case study material as well as a number of research tasks and examination-style questions – some of which could utilise information derived from the 'Recommended Websites' on page 147. Some of the case study material in the book is based on Britain and Brazil. For this reason, a table on page 3 provides a wide range of comparative data for these two countries, among others quoted in the book – together with equivalent information for the world as a whole.

Data comparison table

Data type	Units	World	Botswana	Egypt	Nigeria	China	India	Japan	Russia	Saudi Arabia	Australia	Indonesia	France	Germany	Sweden	UK	Brazil	Colombia	Mexico	Canada	USA
Total population	millions	6,379.0	1.6	76.1	137.3	1298.9	1065.0	127.3	143.8	25.8	19.9	238.5	60.4	82.4	9.0	60.3	184.1	42.3	105.0	32.5	293.0
Population growth rate	%	1.14	-0.9	1.8	2.5	0.6	1.4	0.1	-0.5	2.4	0.9	1.5	0.4	0.0	0.2	0.3	1.1	1.5	1.2	0.9	0.9
Population aged 0–14 years	%	28.2	39.2	33.4	43.4	22.3	31.7	14.3	15.0	38.3	20.1	29.4	18.5	14.7	17.5	18.0	26.6	31.0	31.6	18.2	20.8
Population aged 15–64 years	%	64.5	56.2	62.2	53.7	70.3	63.5	66.7	71.3	59.3	67.2	65.5	65.1	67.0	65.2	66.3	67.6	63.9	62.9	68.7	66.9
Population aged 65 and over	%	7.2	4.6	4.3	2.9	7.5	4.8	19.0	13.7	2.3	12.8	5.1	16.4	18.3	17.3	15.7	5.8	5.0	5.5	13.0	12.4
Median population age	years	27.6	19.2	23.4	18.1	31.8	24.4	42.3	37.9	21.2	36.3	26.1	38.6	41.7	40.3	38.7	27.4	25.8	24.6	38.2	36.0
Fertility rate	children born/woman	2.6	3.2	3.0	5.3	1.7	2.9	1.4	1.3	4.1	1.8	2.5	1.9	1.4	1.7	1.7	2.0	2.6	2.5	1.6	2.1
Birth rate	births/1000 people	20.2	24.7	23.8	38.2	13.0	22.8	9.6	9.6	29.7	12.4	21.1	12.3	8.5	10.5	10.9	17.3	21.2	21.4	10.9	14.1
Death rate	deaths/1000 people	8.9	33.6	5.3	14.0	6.9	8.4	8.8	15.2	2.7	7.4	6.3	9.1	10.4	10.4	10.2	6.1	5.6	4.7	7.7	8.3
Infant mortality rate	deaths/1000 live births	50.3	70.0	33.9	70.5	25.3	57.9	3.3	17.0	13.7	4.8	36.8	4.3	4.2	2.8	5.2	30.7	21.7	21.7	4.8	6.6
Net migration rate	migrants/1000 inhabitants	–	0.0	-0.2	0.3	-0.4	-0.1	0.0	1.0	-2.7	4.0	0.0	0.7	2.2	1.7	2.2	-0.3	-0.3	-4.9	6.0	3.4
Life expectancy	years	64.0	31.0	70.7	50.5	72.0	64.0	81.0	66.4	75.2	80.3	69.3	79.4	78.5	80.3	78.3	71.4	71.4	75.0	80.0	77.4
Population living below poverty line	%	–	47.0	16.7	60.0	10.0	25.0	N.A.	25.0	N.A.	N.A.	27.0	6.5	N.A.	N.A	17.0	22.0	55.0	40.0	N.A.	12.0
Literacy rate	%	77	80.0	57.7	68.0	91.0	59.5	99.0	99.6	78.8	100.0	87.9	99.0	99.0	99.0	99.0	86.4	92.5	92.2	97.0	97.0
Population working in agriculture	%	–	N.A.	32.0	70.0	50.0	60.0	5.0	12.3	12.0	5.0	45.0	4.1	2.8	2.0	1.0	23.0	30.0	18.0	3.0	2.4
Population working in manufacturing	%	–	N.A.	17.0	10.0	22.0	17.0	25.0	22.7	25.0	22.0	16.0	24.4	33.4	24.0	25.0	24.0	24.0	24.0	23.0	24.1
Population working in service industries	%	–	N.A.	51.0	20.0	28.0	23.0	70.0	65.0	63.0	73.0	39.0	71.5	63.8	74.0	74.0	53.0	46.0	58.0	74.0	75.9
GDP per capita	$	8,200	9,000	4,000	900	5,000	2,900	28,200	8,900	11,800	29,000	3,200	27,600	27,600	26,800	27,700	7,600	6,300	9,000	29,800	37,800
Inflation rate	%	–	9.2	4.3	13.8	1.2	3.8	-0.3	13.7	0.5	2.8	6.6	2.1	1.1	1.9	14	14.7	7.1	4.5	2.8	2.3

1 Population Change and Control

Birth control contraception, sterilisation and abortion techniques used to reduce birth rates

Birth rate number of live births in a year per 1000 births

Census survey designed primarily to discover the size of a population

Death rate number of deaths in a year per 1000 people

Demographic Transition Model graph showing how changing birth and death rates tend to be linked to a country's economic development

Dependency ratio relationship between the numbers of people within and outside the economically productive age range (15–65 years)

Fertility ratio the number of births in a year per 1000 women in the child-bearing age range (about 15–50 years)

Globalisation processes and trends operating on a global (worldwide) scale

Less economically developed countries (LEDCs) the world's poorer countries

More economically developed countries (MEDCs) the world's richer, highly-industrialised countries

Natural increase difference between a country's birth and death rates; population change due to migration is *not* a factor in natural increase

Neo-Malthusianism recently increased support for the 200-year-old belief of Thomas Malthus that periods of population growth are inevitably followed by war, famine and other checks on further growth

Population change how the population of an area changes due to birth rate, death rate *and* migration factors

Population doubling time number of years during which the world's population doubles in size

Population policy strategy adopted by a country in order to influence the future growth of its population

Population pyramid horizontal-bar diagram used to display population sub-divisions according to sex and age ranges

Population structure the composition of a population in terms of age and sex

United Nations Fund for Population Activities (UNFPA) agency established in 1969 to support countries' population programmes with respect to family planning, maternal/child health care, educational provision, and the formulation, implementation and evaluation of population policies

1 Introduction: Should We Believe Malthus?

This chapter is probably the most important of all those in the book. It considers the nature and effects of population change on a planet with limited resources, as well as the many factors that determine human reproduction – the **birth rate** component in the formula below:

Population = Birth rate ± Death rate ± Migration
change (Chapter 1) (Chapter 2) (Chapter 3)

The final chapter of the book investigates population distribution, one of the chief consequences of population change.

The Reverend Thomas Malthus (1766–1834) had a very enquiring mind and his concerns about the world's capacity to sustain continued population growth led him to publish his *Essay on the Principle of Population* in 1798. In this essay, he asserted that human population tends to increase in a **geometric** (or **exponential**) **sequence** (i.e. 1, 2, 4, 8, 16, 32, etc.), whereas food production rises are much more likely to take place in an **arithmetic sequence** (1, 2, 3, 4, 5, 6, etc.). He did recognise that improved technology would lead to some improved food **yields**, but he was also convinced that population control factors such as war, disease and famine would begin to take effect soon after an **optimum population** level was exceeded. Malthus acknowledged that birth control (contraception) methods might also have a role to play in limiting population increase, but was convinced that their effect would be minimal; he actually believed such methods to be 'vices', activities that should not really be encouraged!

Only three years later, Malthus revised his initial population and food increase prediction rates by reducing his previous emphasis on the geometric sequence for population increase. However, he remained extremely **pessimistic** about the population–resource balance, a view that achieved renewed support in the second half of the twentieth century as a result of the global **population explosion** triggered by widespread, high **natural increase** rates in many **less economically developed countries (LEDCs)**. This later rise in support for his beliefs is known as **neo-Malthusianism**. In the mid-1960s, a US biologist, Paul Ehrlich, began the modern ecological movement's resurrection of Malthusian thinking. In his book *The Population Bomb* (1968), Ehrlich predicted that hundreds of millions of people would die of starvation in the 1970s, but was not convinced by evidence to the contrary and even increased his prediction by forecasting the starvation of billions in the 1980s. In 1968, a group of 30 scientists, economists and other specialists met in Rome to discuss what they believed to 'the future predicament of man' and later formed what became as the **Club of Rome**. Their early discussions focused on five basic factors that are capable of influencing global development: population, agricultural production, natural resources, industrial production and pollution. The group concluded that population pressures had already reached such a high level that humankind had no option but to seek a 'state of equilibrium'. One of their main justifications was the decreasing world **population doubling time**; another was their belief that technological innovation alone could not make it possible to support such a high and increasing rate of population increase.

Esther Boserup was one of many **optimists** who had little or no faith in the Malthusian theory – even its revised 1801 version! In 1965, she argued that *all* people have the will-power and potential resources of knowledge and technology to achieve adequate increases in food production. She therefore viewed population growth in pre-industrial societies as a **positive factor** – one that would lead to an increase in food production that would more than compensate for any net population increase. Boserup's philosophy was based on her observations of communities at different **levels of development**. She believed that people had an intuitive awareness of more effective, **intensive farming** methods that they could put into practice in times of need.

Others, including a French sociologist called Durkheim, have become convinced that population growth *inevitably* leads to an increase in food production. Their argument is that population increase results in a greater sub-division of labour, in which the farming community becomes smaller but much more efficient, thus allowing others to specialise in other, non-agricultural activities. Other influential optimists included the Dane Bjorn Lomborg, whose book *The Sceptical Environmentalist* (written after a series of newspaper articles printed in 1998) pointed out that, far from increasing starvation, population trends had been accompanied by reduced rates of global poverty and increased life expectancy; also the US economist Julian Simon, whose book *The Ultimate Resource* (1980) proposed the argument that human ingenuity was perfectly capable of overcoming short-term resource deficiencies through exploration, re-cycling and invention.

Section 4 investigates the population–resources balance issue in some depth.

2 The Role of the Census

A **census** is a survey, whose primary function is to obtain an up-to-date total of the population of a certain area. The first census of England was carried out just over 200 years ago – but only after a series of bitter debates in parliament. One of the chief concerns of the Members of Parliament at that time was that a census might show a smaller population than predicted, which could make potential invaders more confident of success. Many other MPs were in favour of having censuses, at regular intervals, arguing that it was important for the government to know how quickly the population was increasing. A number of other European countries had already established censuses, for example Sweden in 1748 and Austria in 1754.

The first census in England was carried out on 10 March 1801 and indicated that its total population at that time was 9,168,000 people. However, this was very likely to be an underestimate, because the **enumeration** (the census counting process) was supposed to be undertaken by local parish officials, but many of them simply could not be bothered to take part. The counting process was extremely basic and it was not until the 1821 census that information about a person's age was asked for – it had previously been considered far too tactless to enquire about such 'personal' information. Even then, only an approximate estimate of age was requested. The 1841 census had been made considerably more effective by an Act of Parliament that made the registration of all births, marriages and deaths compulsory from 1 July 1837, but it was not until 1851 that a census question was included requiring a precise 'age at last birthday'. More recent censuses have demanded a much greater range of information, including details of occupation, car ownership and country of birth. Many additional questions have since been added to provide information on current issues of concern. For example, the 1891 census requested details of the number of rooms within each household because of fears about overcrowding in rapidly expanding industrial towns and cities. The 1991 census was the first to include a question about **ethnicity** and the 2001 census sought information about the amount of time spent caring for people with long-term physical or mental disabilities, reflecting growing concern about the UK's ageing population and the need to plan for increased care provision. Additional information about ethnic origins was also required by the 2001 census, partly to provide statistical information about migration patterns. A national census has been held every 10 years since 1801, except in 1941 when the Second World War was at a critical stage.

Census data can prompt interesting questions as well as confirm suspected trends. Recognising this fact, the Soviet Union deliberately avoided holding a census for many years after the Second World War, because it feared what everyone already knew: the war had not only resulted in a huge loss of life, but also led to a serious imbalance in the country's adult male to female population ratio. In 2002, Russia undertook its first census since the collapse of the Soviet Union and the fall of the Berlin Wall in 1989. The results published a year later confirmed that the Russian population had continued to decrease. This more recent decline is, however, as a result not of human conflict, but AIDS, increased alcoholism and a general deterioration in that nation's health.

The UK's 2001 census revealed a national population total of 58,789,194 and a net growth rate of 4.1% over the previous 10 years. It also indicated that a very modest growth rate for the **indigenous population** had taken place and that international migration was an increasingly significant component in that overall increase. The Registrar General for England and Wales was quite surprised that the 2001 census had also revealed a shortfall of about 900,000 people when compared with the pre-census estimates. The Registrar General speculated that much of this unpredicted population 'loss' was due to what appeared to be a new demographic phenomenon: an exodus of young adults, of students in general and males in particular, within the 25–29-year age range, who had taken their gap years overseas or been enticed abroad by the Mediterranean region's 'rave culture'. The Registrar General also announced that the £240 million cost of the previous 10-year census cycle had been fully justified, because it would provide up-to-date statistics that would form the basis of future planning in educational, medical, transportation and other crucial areas of government spending. The Registrar General did, however, acknowledge a number of difficulties in trying to achieve data consistency over recent decades, such as the fluctuating 'mobile' populations of hospitals, prisons, universities, boarding schools and military establishments.

Some of the most important statistics and trends revealed by the UK's 2001 census were:

- The UK population consisted of 49.1 million (83.5%) in England, 5 million in Scotland, 2.9 million in Wales and 1.7 million in Northern Ireland.
- Of all four UK countries, only Scotland's population showed a decrease, a 2% fall between 1981 and 2001.
- The region with the fastest-growing region population was south-west England (+12.5% since 1981), followed by the east (+11%) and south-east England (+10%).
- North-east England experienced the highest regional population loss (−5% since 1981), followed by north-west England (−3%).
- Greater London's population continued its post-Second World War rise: up 5.4% since 1981.
- For the first time, people aged 60 or over formed a larger part of the population than children under 16 (21% compared with 20%, respectively).
- There continued to be a substantial increase in the number of people aged 85 and over, now over 1.1 million (1.9% of the total population).
- Christchurch, in Dorset, had the highest proportion of people of retirement age at 33.1%
- Eastbourne, in Sussex, had the lowest proportion of adult males, 87 per 100 women.

CASE STUDY: CENSUS ENUMERATION DISTRICTS IN LEYLAND

Although once world-famous for manufacturing cars, buses and lorries, the main function of the central Lancashire town of Leyland for most of its existence was providing essential services for the farming communities on its surrounding fertile plain. The town's agricultural past is still recalled in street names such as Fox Lane, Broadfield Drive and Golden Hill Lane (named after the corn which once grew there).

Industrialisation came late to Leyland, mainly because most of the land that is now developed used to be owned by two wealthy families that did not like parting with it. Leyland was linked to the national railway network in 1838, an event that prompted the building of some cotton-weaving mills on the western side of the town, as well as some streets of the terraced housing characteristic of that period. In 1895, the Lancashire Steam Motor Company was established near the town centre. At first it specialised in making lawn mowers and steam-powered road vehicles. Rapid expansion followed the development of a reliable petrol-driven engine and the company later moved to a much larger site on the northern edge of the town. Much of Leyland's current housing stock dates from the 1920s and 1930s, when the motor works expanded yet again. The town later built a number of council housing estates.

The most recent phase in the town's development began in 1971, following parliamentary approval for Leyland and nearby Preston and Chorley to be the growth points for the Central Lancashire New Town development. This decision led to the building of many new factories and houses, mainly in the town's southern and western districts. It also meant that Leyland's population increased at a time when most British industrial towns were experiencing a population loss. In the 2001 census its population was 37,100.

Figure 1.1 shows the present layout of the town and includes the locations of the five contrasting residential zones examined in the rest of this case study. For census purposes, every town and city in the UK is divided up into wards and enumeration districts. **Wards** are sizeable areas, each with 2000–8000 people; at present, Leyland has six wards. The census information for wards is used by local authorities for planning purposes. This information is also of great interest to geographers, especially when wards are mainly composed of just one type of housing. **Enumeration districts** are much smaller than wards and the five selected for this study have an average population of 580, each of them having only one or two streets. Most of Leyland's housing stock is less than 100 years old and much of it less than 50. The choice of enumeration districts reflects this, as only one of them was in existence before 1900.

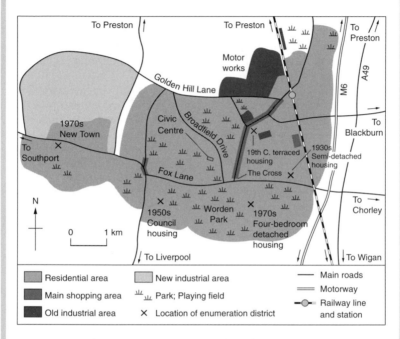

Figure 1.1 Enumeration districts in Leyland

Figure 1.2 provides four important types of information about each of these districts. Information has not been provided on the proportions of 'sub-standard' housing (e.g. not having an inside bathroom or toilet) or the ethnic origins of the population, because both sets of data are relatively insignificant throughout the Leyland urban area. Readers can devise their own preferred means of utilising the information in these graphs; comparison tables are particularly suitable because their data can be accessed so easily by eye.

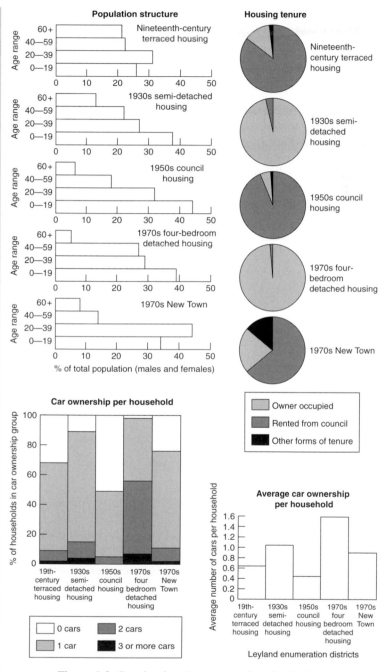

Figure 1.2 Graphs showing census data for Leyland

CASE STUDY: POPULATION CENSUS TRENDS IN BRITAIN

The UK censuses carried out over the past 200 years have provided a wealth of detailed information about the nation's changing population, particularly its increasing number of people. Figure 1.3 illustrates how the population of Britain (England, Scotland and Wales) has changed over the last two millennia, incorporating data information from the 1801 and subsequent censuses. Census information for Ireland is not available for 1801 and 1811; the changing political composition of this island during the past century makes it difficult to standardise collated statistics.

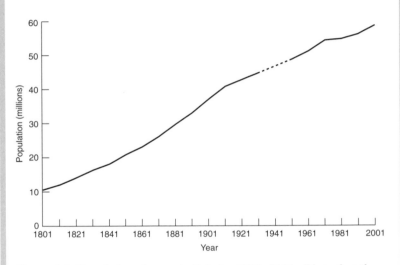

Figure 1.3 Population change in Britain: 1801–2001. (Note that there was no census in 1941 because of the Second World War.)

Figure 1.4 shows population change, as a series of percentage differences for each 10-year census interval. Section 1 explained that changes to population totals are the net result of three variables: birth rates, death rates and migration. Figure 1.5 illustrates recent changes in the first two variables, the most important of these variables.

Inter-census period	Population change over inter-census period (%)
1801–11	14.3
1811–21	17.5
1821–31	15.6
1831–41	11.0
1841–51	7.2
1851–61	11.1
1861–71	13.0
1871–81	13.8
1881–91	11.1
1891–1901	12.1
1901–11	10.3
1911–21	4.9
1921–31	4.7
1931–51	4.5*
1951–61	4.9
1961–71	6.3
1971–81	0.7
1981–91	2.6
1991–2001	3.8

* Both 10-year percentage averages based on the 1931 and 1951 census figures.

Figure 1.4 Inter-census population changes in Britain: 1801–2001

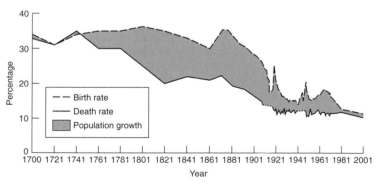

Figure 1.5 Britain's changing birth, death and population growth rates: 1700–2001

CASE STUDY: MORTALITY TRENDS SURVEY BASED ON ST MARY'S CHURCHYARD, BARTON

St Mary's Churchyard lies 7 km north of Preston, in central Lancashire, and a recent survey of its burial headstones forms the basis of the information in Figure 1.7. The 20-year average figures in the last column are shown in Figure 1.6. The most important features highlighted by the table and this graph are:

- there is an *overall* increase in longevity between 1800 and 2000
- there was a *continuous* increase in longevity throughout the twentieth century
- the greatest two-decade increase in longevity (of 16 years) took place in 1800–20
- the sixth column reveals a number of significant fluctuations in the relative mortality of males and females. The unusually high figure for the 1840s–50s (showing males having a 12-year greater longevity than females) may be explained by the unreliability of trends based on the very modest number of burials for that period
- there was a reduced number of burials in the 1980s–90s. This was at a time when the populations of Barton, its neighbouring villages and the countryside between them were all increasing due to **counter-urbanisation**, chiefly from nearby Preston. One possible explanation for the reduced number of burials recorded for that period was the Roman Catholic Church's decision, in 1983, to regard cremation as an acceptable alternative to burial.

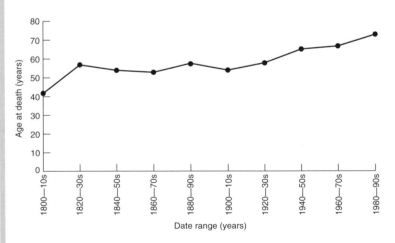

Figure 1.6 Changes in mean age at death: based on survey of Barton churchyard

Survey period	No. of males	Average age of males	No. of females	Average age of females	Female average ±	Total no. of all males and females	Average age of all males and females
1980–90	41	72	39	73	1	80	73
1960–70	73	65	69	70	5	142	67
1940–50	64	63	71	67	4	135	65
1920–30	41	58	49	58	0	90	58
1900–10	39	55	48	53	−2	87	54
1880–90	37	59	44	58	−1	71	58
1860–70	42	50	31	56	6	73	53
1840–50	14	60	15	48	−12	29	54
1820–30	9	56	12	60	4	21	57
1800–10	11	41	13	41	0	24	41

Figure 1.7 Table of survey information for Barton churchyard

CASE STUDY: POPULATION CENSUS TRENDS IN BRAZIL

Figures 1.8 and 1.9 illustrate Brazil's changing total population and birth and death rates. Brazil's population statistics for the nineteenth century are much less reliable than those for Britain. Even so, there are very clear differences in the trends for both countries. The total population trends accelerate at different periods of time and at very different rates. The birth and death rates also display some significant differences, but the patterns do share some similarities, although these also appear to have taken place at quite different periods of time. The significance of this 'time-shift' is discussed in the next section.

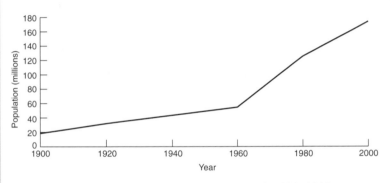

Figure 1.8 Population change in Brazil: 1900–2000

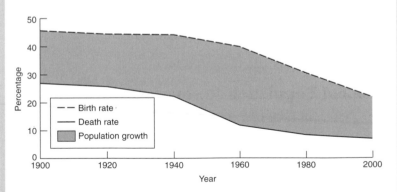

Figure 1.9 Brazil's changing birth, death and population growth rates: 1900–2000

3 Demographic Transition

The **Demographic Transition Model** (Figure 1.10) is well named, because its three component words have the following meanings:

- 'demographic' means that population trends are being considered, in this case birth rates, death rates and the net result: population increase/decrease
- 'transition' means that significant changes are taking place over a period of time
- 'model' is the general term used to describe a theoretical way of illustrating what often occurs in the real world.

This particular type of model is designed to show how certain key population rates change over time. It was based initially on population trends in Western Europe, because most of those countries had long-established census programmes capable of providing accurate and reliable statistical data. Analysis of data gathered in this way revealed some quite distinct patterns of population rate change; also that these changes appeared to be closely linked to the rate at which individual countries develop (i.e. became more wealthy, giving the majority of their citizens an improving level of **quality of life**). Later comparisons with less developed countries in other continents supported this theoretical link between a nation's wealth and its key population change rates. It should now be clear why the last section ended with comparative population data for Britain and Brazil: the trends are very similar, the only main difference being the 'time-shift' due to the relative wealth of the two countries at different periods.

Figure 1.10 is quite detailed, because it is important not only to recognise the population trend characteristics of each of the model's five phases, but also to understand *why* each phase is so different. The global significance of the Demographic Transition Model is considerable; by knowing which countries are likely to be at each phase, it is possible to make realistic predictions about population growth of the world as a whole.

4 Global Population Resources

a) Non-renewable resources

In some ways Malthus was perfectly correct to be concerned about the world's limited **carrying capacity** – its ability to sustain a constantly increasing global population. Most of the Earth's **natural resources** are **finite** (i.e. not limitless). One prime example of this is the small range of **fossil fuels** – coal, oil (petroleum) and natural gas – whose known reserves seem unlikely to extend much beyond about 2240, 2050 and 2075, respectively, *at the current rates of usage*. These important sources of global energy are also the raw materials used to produce a

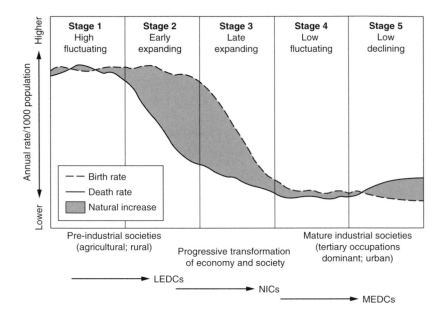

UK phases	Pre 1750	1750–1870	1870–1940	Post 1940	Not yet reached
Brazil phases	Pre 1890	1890–1960	1960–7	Not yet reached	
Birth rate characteristics	High fluctuating	High small decrease	Rapid decrease	Low fluctuating	Low
Death rate characteristics	High fluctuating	Rapid decrease	Less rapid decrease	Low fluctuating	Very low
Natural increase characteristics	Low variable	Rapid increase	Less rapid increase	Low variable	Small decrease
Characteristics factors	Poor sanitation; frequent epidemics; no pensions or care facilities; children economically important	⟶ infant mortality decreases/longevity increases ⟶ ⟶ IMPROVED QUALITY OF LIFE ⟶ • availability of contraception • educational standards • food production and quality • housing provision • medical knowledge • personal affluence • sanitation and water purity • smaller families			
Exemplar countries	–	Bangladesh Ethiopia Zaire	Brazil China India	Canada UK USA	Germany Italy Sweden

Figure 1.10 The Demographic Transition Model

wide range of manufactured goods including cosmetics, detergents, dyes, explosives, fertilisers, medicines, synthetic fibres and tar (for road surfaces). Our modern way of life would be extremely difficult to sustain without these.

(i) Wood

Wood is another common source of energy, particularly in LEDCs, and is a potentially **renewable resource** because felled trees can be replaced by **re-afforestation** programmes. Metal ores such as iron ore and bauxite (from which aluminium is obtained) are most certainly finite, hence the increasing awareness of the need for recycling measures, especially in countries such as Japan which are heavily reliant on imports of industrial raw materials. Soil is obviously a vital natural resource, one which can be permanently lost due to erosion and can quickly become infertile if not protected adequately and is discussed further in Chapter 4.

(ii) Water

The most essential of all natural resources is water, whose total quantity within the Earth's **hydrosphere** never alters, although its locations and forms (ice, liquid and vapour) are constantly subject to change. Figure 1.11 shows the current proportionate forms of the total volume of 'water' within the hydrosphere.

(iii) Food

Food is without doubt humankind's second most important basic resource. At the time that Malthus was stimulating interest in demographic matters, there was very limited knowledge about the total number of people who inhabited the planet. Significantly, his essay was published only three years before the first British census took

Hydrosphere component	'Water' volume (millions of cubic kilometres)	Percentage of total volume
Atmosphere	0.013	0.00094
Biosphere	0.0006	0.00004
Groundwater flow	8.32	0.59960
Ice caps and glaciers	29.0	2.08994
Lakes	0.16	0.01153
Oceans and seas	1350.0	97.29031
Rivers	0.04	0.00288
Soil moisture	0.066	0.00476
Total	1387.5996	100.00000

Figure 1.11 Components of the hydrosphere

place. One of the greatest weaknesses of Malthus' theory is that he underestimated the human capacity for increasing food supplies in times of need. This is well illustrated by the following summary of the progression from personal self-sufficiency in food resources to the **high-tech, mass-production** technologies of today.

b) Self-sufficiency

The earliest communities had to be completely **self-sufficient** in order to survive. There are now very few remnants of totally self-sufficient communities, most of the remaining examples being numerically very small and still contracting as more of their members improve their quality of life by adopting 'modern' ways of supporting themselves. The Aborigines of the Australian outback are a case in point. It is their acquired understanding of their surroundings and the way in which their tribal community is organised that have enabled them to become so self-sufficient in a hostile natural environment. They traditionally used spears tipped with sharp stones to kill kangaroos and fish; their boomerang invention was ideal for killing birds because it returned to the thrower after failing to strike the target – so minimising any wastage of scarce raw materials. Animal life was never killed for sport, only to provide food. Plant regeneration was achieved by controlled burning; each tribe had its own territory, whose boundaries and natural resources were deeply respected by neighbouring tribes.

c) Early agricultural development

Britain's Neolithic people lived between about 25,000 and 4000 BCE (before common era), when the total population was so small that it might just fill a 50,000-seat football stadium. As the number of people and rate of urbanisation both increased, it became necessary for more intensive farming methods to be devised. By about 800 CE (common era), the best arable land was divided up into strips, separated by grass-covered strips called baulks that acted as boundaries between each farmer's plots. A three-field **crop rotation** system operated in the small fields, which involved planting wheat in the first year, barley in the second, then leaving the land **fallow** (resting) in the final year to self-restore its fertility. After crop harvesting, animals were allowed to graze on the remaining stubble in order to accelerate this process.

d) The first agricultural revolution

Britain's first agricultural revolution took place in the eighteenth century, and transformed both arable and pastoral farming. Fields were enclosed by walls and hedges, and the urgency to increase food yields led to improved technologies such as those devised by the three pioneering farmers below:

- Robert Bakewell, who improved the breeding of animals. His 'New Leicester' sheep carried more flesh and provided much needed meat for Britain's growing urban population.
- Viscount Thomas Townshend, a wealthy landowner, who considered it wasteful to leave land fallow every third year. He developed a new, four-year, rotation using root crops to maintain soil fertility (Figure 1.12). He quickly acquired the nick-name 'Turnip Townsend'.
- Jethro Tull, who invented a seed-drill that was much more efficient than the previous technique of dispersing seeds by hand.

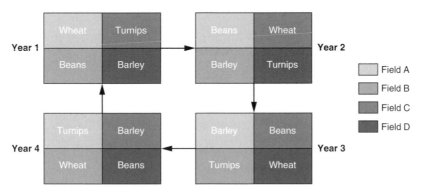

Figure 1.12 The four-year crop rotation system

e) Plantation farming

One agricultural system that operated very effectively for a number of centuries is the **plantation system**. Plantations are very large farming units, usually in tropical and semi-tropical regions and specialising in one major **cash crop**. The first plantations grew sugar cane and were laid out in Brazil and islands such as Jamaica in the West Indies during the sixteenth century. Other major plantation crops were: cacao (for making chocolate products), coffee, rubber, tea and tobacco. The many advantages of the plantation system have resulted in much heavier crop yields. Some of the most important of these advantages are:

- plantation workers become highly experienced in growing one or at most two crops
- large plantations can finance research into crop disease prevention
- large plantations can afford to buy the most up-to-date crop processing machinery, thus greatly reducing onward transportation costs.

Plantations do have some disadvantages, especially those that risk using **monoculture** (specialising in just one crop), making them vulnerable to pests and plant diseases. Their great advantage is that they can produce large quantities of food and other raw materials efficiently and cheaply, they use the land intensively and they do create additional protection against soil erosion.

f) The second agricultural revolution

Farming is now as competitive as any other major industry in Britain and some of the largest farms are owned by companies wishing to **diversify** from their main manufacturing and investment activities. The nature of Britain's second (twentieth-century) agricultural revolution has prompted a very appropriate term: **agribusiness**. Some of its key features are:

- the increased use of farm machinery, which has allowed the workforce to be reduced accordingly (Figure 1.13)
- the use of much larger machinery, which led to hedgerows being grubbed-up and ponds filled-in to enlarge fields and enable such machinery to operate more easily
- 'battery farming', to increase the yield of chickens and eggs
- 'artificial insemination', which allows selective stock breeding to be undertaken, using only high-grade bulls for breeding purposes

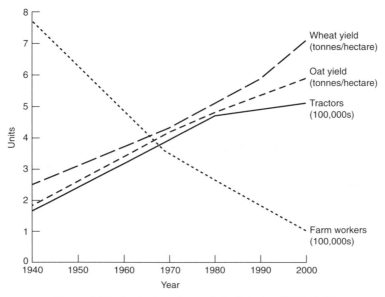

Figure 1.13 Agricultural trends in Britain: 1840–2000

- the widespread use of increasingly sophisticated chemical applications, the chief types being:
  - fertilisers, to increase the volume of essential plant nutrients in the soil
  - fungicides, to combat plant diseases
  - herbicides, to kill grasses and other competitor weeds
  - pesticides, to kill insects which eat or otherwise damage commercial crops.

g) The Common Agricultural Policy

The European Union (EU) was formed in 1957 with six founder states. The UK joined in 1973 and by 2004 the EU's membership had increased to 25. Two of the main initial aims of the EU were to make its farmers more efficient producers of food and raise their standard of living. Its strategy for achieving these aims was the **Common Agricultural Policy (CAP)**, a system that guaranteed the prices at which farmers could sell their produce and ensured that *all* the food they produced would be bought in this way. The scheme proved so successful that food surpluses became embarrassingly large and the cost of storing them prohibitive. During the 1970s and 1980s, Britain alone needed 100 chilled **intervention stores** for its excess 50,000 tonnes of beef and 250,000 tonnes of butter. Across the whole EU, there was enough wine in storage to fill 700 Olympic-size swimming pools. The cost of keeping these 'food mountains' and 'wine lakes' totalled £200 million *every week*, which represented half of the EU's entire annual budget. Some of this surplus food was donated to needy people and 250,000 tonnes of butter was issued to Britain's poor and elderly early in 1987. However, relatively little was exported at subsidised rates, except as emergency aid in times of extreme famine. The reason most often given for this apparent meanness is that dumping large quantities of heavily subsidised food can undermine the local agricultural economy, simply by making it not worthwhile for the local farmers to continue producing food.

h) Other advances in food production technology

Other advances that Malthus could not have foreseen include the following:

- Large-scale draining of wetlands and reclamation of shallow coastal areas to create level, highly fertile arable land, even though the Fens and Somerset Levels had already been subjected to drainage schemes since the 1750s.
- Exploitation of mid-latitude grasslands in North America, Eurasia and Australia for extensive ranching and cereal cultivation.
- Selective breeding of higher-yield crop strains, especially rice, that was one of the main features of the **green revolution** which took

place in the 1950s and 1960s in many LEDCs such as India; other green revolution strategies included local irrigation systems, better professional training for farmers, the availability of bank loans for agricultural improvements and reforms to consolidate fragmented parcels of land.

- Using greenhouses to protect crops from adverse weather and 'force' them to ripen faster; in Japan, farmers have become expert in growing vegetables indoors under computer-controlled conditions and have even developed ways of growing cube-shaped water melons which fit into fridges more easily!
- Using extensive irrigation systems that have transformed fruit and vegetable production in areas like California's Central Valley, which uses water from the Upper Sacramento Basin.
- Great improvements in transportation and food preservation and storage (e.g. refrigerated cargo ships and air freight for the speedy carriage of perishable foods; the chilling, freezing and dry-freezing of foodstuffs).

5 Changing Global Fertility Rates

Figure 1.14 traces the world's changing population and the three very different phases of change that have already taken place. The most notable of these phases was the so-called population explosion, the result of many of the largest LEDCs being at phase 2 of the Demographic Transition Model, in which birth rates exceeded death rates for the many reasons discussed earlier. Some of the most heavily populated of these countries have now not only introduced effective **population control** programmes (China and India being the outstanding examples), but have also developed economically to the point where they have progressed to phase 3, with its much reduced birth, death and – most importantly – population growth rates. Many countries still remain within phase 2, unable to kick-start their national economies sufficiently to escape the cycle of poverty (Figure 1.15) in which they appear to be trapped. The populations of some states (especially in Africa) have declined significantly, but even their collective influence has not been enough to offset the changing global population trends dominated by the larger, increasingly developed countries.

It is known that families are now having far fewer children. During the 1970s, global fertility rates stood at about six children per woman, whereas today's average is only 2.9, and is beginning to approach the critical rate of 2.1 at which human populations merely replace themselves. The world's population now appears likely to rise to about nine billion by 2050 (an increase of 50% on the present figure), but then decrease rapidly during the second half of the century on a scale not seen since the Great Plague (the 'Black Death') of 1348. It is also

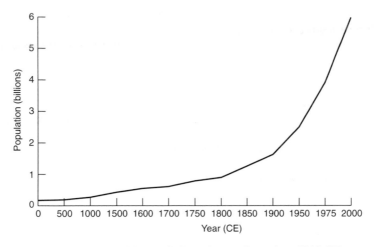

Figure 1.14 World population change from 0 to 2000 CE

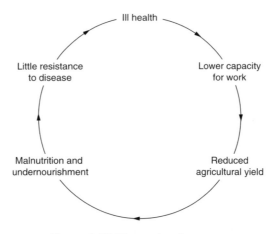

Figure 1.15 The cycle of poverty

widely accepted that increasing prosperity discourages childbearing and many European countries provide good examples of this. Europe's overall birth rate of 1.4 is well below the critical population replacement rate. There are significant regional variations, from 1.8 per woman in France and Ireland to 1.2 in Italy and Germany. Britain's current fertility rate is 1.7. Only Albania and Kosovo are not following the overall European trend towards reduced fertility rates. It now seems inevitable that Western Europe's population will decrease by several million a year as today's young adults reach old age, die and are not replaced. Other countries with birth rates below the replacement fertility rate level include Australia, Sri Lanka, Cuba,

Uruguay, Brazil, Mexico, and the rapidly developing 'tiger economy' nations of south-east Asia such as Singapore and Taiwan.

Some countries are now experiencing a particularly serious population decline. Russia's population is currently decreasing by almost 750,000 per year, a phenomenon appropriately described by President Putin as 'a national crisis'. The decline's causes are alcoholism, a breakdown of the public health system and industrial pollution that appears to have had a disastrous effect on the sperm counts of Russian men. China's reducing fertility is due to its adherence to a 'one-child' policy, which will be discussed in the next case study. China's population is expected to peak at 1.5 billion by 2019, then enter a steep decline. Some analysts suggest that China could lose 20–30% of its population *every generation*. Already, most young Chinese people have no brothers or sisters and face the prospect of having to care for two parents and four grandparents by themselves; the country's incomes and pensions are simply not rising quickly enough to support an ageing population on the scale predicted by demographers.

On the other hand, there are still many nations – mainly in Africa and the Middle East – where fertility rates remain high. According to the UN, the population of the Middle East could double in the next 20 years. Saudi Arabia has a fertility rate of 5.7; in Yemen, it is 7.2 and in Palestine 5.9. More surprising is the case of Africa. Despite AIDS, which kills millions of young Africans every year, that continent's overall population continues to rise – except for South Africa and a few Mediterranean coast nations such as Libya.

Rural–urban migration (discussed in considerable detail in Chapter 3) is another important factor in fertility reduction. The UN recently predicted that half the world's population will live in urban areas by 2007. Because most urban children are dependent on, instead of contributing to, family budgets, there is a clear financial incentive for town-dwellers to consider having fewer children, whereas children in rural areas often continue to play an important traditional role in assisting with farm work.

Population policies have been developed by many countries to influence their birth rates and hence their net growth rates. China's one-child policy is the best known of all the **anti-natalist policies** intended to reduce birth rates, but other countries such as India, Singapore and Sri Lanka have also developed highly effective birth control programmes. Compulsory sterilisations were introduced in India, with financial rewards being offered to women who complied. Singapore provided family planning services and publicised the benefits of having smaller families. Sri Lanka was one of the first countries to recommend the use of contraceptives (locally called *preethi*). Some European countries such as Germany and Sweden are experiencing population losses and are able to offer substantial financial incentives to stimulate higher fertility rates.

CASE STUDY: CHINA'S ANTI-NATALIST POLICY

Shanghai relaxes China's anti-natalist (one-child) policy
(**The Guardian** *14 April 2004*)

As the most crowded city in the world's most populous country, Shanghai is perhaps an unlikely starting place for any relaxation of China's controversial **one-child policy**. But that is exactly what happened yesterday when Shanghai Municipal Government announced that divorcees who re-marry in the city would no longer be penalised for having a second baby. Experts believe that the easing of restrictions in Shanghai is likely to foreshadow similar steps across the whole country, when other local authorities begin to realise that the growing financial and social costs of birth quotas could soon outweigh their benefits.

Taking advantage of a new central government initiative to give localities more freedom to apply family planning policies, Shanghai will in future allow couples to produce a child even if both partners already have one from a previous marriage. Until now, divorcees were only allowed to have a second child if their new spouse was childless. Those who broke this rule were forced to pay 'social compensation' – a fine of up to *three times the total annual household income.*

Xia Yi, the vice-director of the Shanghai Municipal Population and Family Planning Commission did, however, stress that the national one-child policy would still remain the basis for any new regulations. He added that the new rule aims to resolve the growing imbalance between the affluent urban residents (whose numbers seem certain to decline because they obey the rules and tend to have children much later in life) and the rural population, where family planning rules are less rigorously enforced. In Shanghai (population 17 million), the authorities are concerned that the shrinking pool of higher-rate taxpayers will be unable to support an ever-increasing elderly population. China now has 88.1 million people aged 65 or over, equivalent to about 7% of the total population. By 2050, this is predicted to exceed 300 million (25% of the total). Policymakers have expressed alarm about the rising gender imbalance as more couples resort to abortion and infanticide to ensure that their single permitted child is a male – someone who can carry on the family name and undertake manual farmwork. Nationally, there are 115 boys born for every 100 girls, while in some provinces, the ratio exceeds 13:10.

Demographic experts are confident that Shanghai – one of the first cities in China to adopt a one-child policy in the late 1960s – will now be a pioneer in gradually relaxing its potentially damaging natal restrictions.

6 Population Structure and Age Dependency

a) Population structure

Much of this chapter has been concerned with changing population *totals*, but some sections have touched on another very important aspect of demography: **population structure**. This is the **composition** of a population, the way it is sub-divided in certain ways, usually according to age and sex. Such information is best displayed as **population pyramids** (also known as **age–sex pyramids**), pairs of horizontal bar graphs sharing the same age scale. Population structure information is very useful in predicting the future need for facilities such as schools and residential care homes; it also enables governments to forecast what proportion of the population is likely to be available to pay taxes for such facilities.

The examples of population pyramids in Figure 1.16(a) are for the UK, but are typical of most graphs of this type because their vertical scales are divided into five-year age ranges called **quinquennials**. Also, their left-side female and opposite side male population scales are based on percentages, not total numbers of people. Both examples have been shaded so as to highlight three distinct age groups:

- Childhood (or pre-adulthood) group: aged 0–14 years, although its upper age limit can vary considerably depending on a particular country's educational system and the age at which its young people would normally expect to finish their compulsory education and begin full-time paid employment.
- Adult (or productive adult) group: 15–64 years. This also has a somewhat arbitrary choice of upper age parameter, especially as increased longevity is forcing some of the more developed countries to delay the age at which national pensions start to become payable. Another variable is the widespread practice for pensions to start to become payable at different ages for men and women, reflecting the number of years of additional **life expectancy** for females. This age range is 'productive', not only in the economic sense in that it includes the great majority of wage-earners, but also the biological sense because women are capable of bearing children for the majority of those years. Improved nutrition has, however, resulted in a much earlier start to menstruation in many countries.
- Elderly (or pensionable adult) group: aged 65 years and older. Many of the above comments also refer to this last age group, whose members inevitably become increasingly dependent on national, local community and individual family group resources for their financial and medical support. This is a rapidly-expanding group, because life expectancy is increasing in most countries due to improved quality of life. Japan has recently provided some outstanding examples of increased **longevity** and its current average

life expectancy for Japanese women is the highest in the world, at 84.6 years. Until her death in October 2003 at the age of 116, Mrs Kamato Hongo was one of the world's oldest people. At the other extreme of the global life expectancy spectrum is Botswana, whose average national life expectancy is now only 39 years – 30 years less than it would have been without the AIDS **pandemic** (Figure 1.16b3 and b4). The significance of these figures does not really become apparent until they are compared with those of earlier times. During the Roman Empire, life expectancy was only 22 years – a figure that increased by only another 11 years during the following one and a half millennia. Since 1840, the average life expectancy in the longest-living countries has increased at an average of three months every year. In the UK, all those who reach 100 years of age receive a congratulatory birthday telegram from the Queen – as did her own mother in 2000. This is still a very special tribute to a long life, but one which may become much more commonplace in the foreseeable future.

Figure 1.16(b1 and b2) shows some contrasting population structures within the UK. Figure 1.16(c) displays a typical pyramid for each of the five phases of the Demographic Transition Model.

b) Age dependency

A very simple way of assessing whether a country might be able to support its more dependent people is by using the **dependency ratio** formula:

$$\text{Dependency ratio} = \frac{\text{number of people in both the childhood and elderly age groups}}{\text{number of people in the adult (productive) age group}} \times 100$$

The result is a percentage figure, which can be used to make meaningful comparisons between countries, provided it is known that the two inter-group age boundaries are determined in precisely the same way within the countries being compared. Figure 1.17 provides some idea of the wide range of dependency ratios within today's world. This table also includes selected figures that illustrate the wide range of current **infant mortality rates**. Improved medical care provides an increasingly safe introduction to life in many countries of the world, but there remain countries such as Sierra Leone (in West Africa), where an average of 157 per 1000 babies fail to survive to their first birthday. Infant mortality is now regarded as one of the most accurate indicators of a country's standard of development, because it reflects so many key health factors such as nutrition, safe drinking water, sanitation, medical services, housing and training in basic life skills.

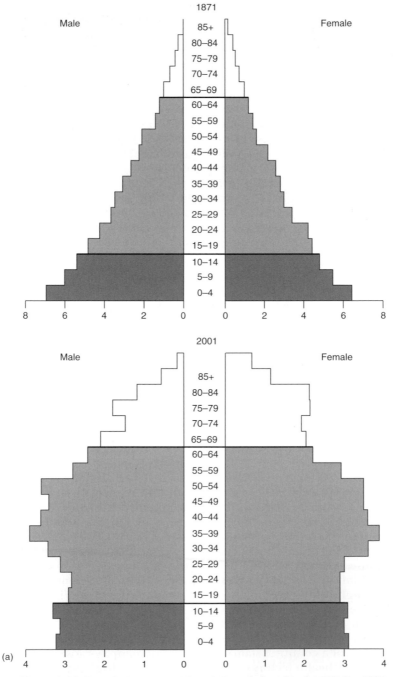

Figure 1.16 Population pyramids. (a) Population for the UK for 1871 and 2000.

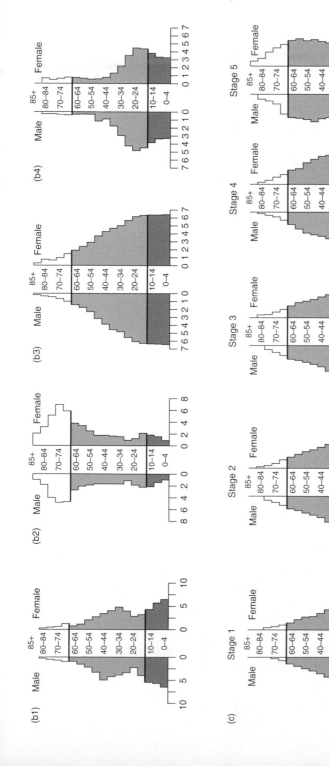

Figure 1.16 (*cont'd*) (b1) Stevenage new town (1966) and (b2) Sussex (2001), a popular retirement region in a south coast resort. (b3) Botswana, prediction for 2020 without any affects for HIV/AIDS and (b4) for 2020 taking these affects into account. (c) Demographic Transition Model stages 1–5.

Country	Age ranges			Dependency ratio	Infant mortality rates (per 1000 live births)
	0–14	15–64	65+		
Burkina	46.0	51.1	2.9	0.96	98.7
Madagascar	44.9	52.0	3.1	0.92	78.5
Mozambique	43.6	53.6	2.8	0.87	137.1
Nigeria	43.4	53.7	2.9	0.86	70.5
Pakistan	40.1	55.8	4.1	0.79	74.4
Zimbabwe	39.4	57.0	3.6	0.75	67.1
Egypt	33.4	62.2	4.4	0.61	33.9
Morocco	32.6	62.5	4.9	0.60	43.3
Mexico	31.6	62.9	5.5	0.59	21.7
Bangladesh	33.5	63.1	3.4	0.58	64.3
India	31.7	63.5	4.8	0.57	57.9
France	18.5	65.1	16.4	0.54	4.3
Sweden	17.5	65.2	17.3	0.53	2.8
Indonesia	29.4	65.5	5.1	0.53	36.8
UK	18.0	66.3	15.7	0.51	5.2
Luxembourg	19.0	66.5	14.5	0.50	4.9
Japan	14.3	66.7	19.0	0.50	3.3
USA	20.8	66.8	12.4	0.50	6.6
Italy	14.0	66.9	19.1	0.49	6.1
Germany	14.7	67.0	18.3	0.49	4.2
Australia	20.0	67.2	12.8	0.49	4.8
Ireland	21.0	67.5	11.5	0.48	5.5
Brazil	26.6	67.6	5.8	0.48	30.7
Netherlands	18.3	67.8	13.9	0.47	5.1
Spain	14.4	68.0	17.6	0.47	4.5
Canada	18.2	68.7	13.1	0.46	4.8
China	22.3	70.2	7.5	0.42	25.3
Russia	15.0	71.3	13.7	0.40	17.0

Figure 1.17 Dependency ratios and infant mortality rates for selected countries

Another very useful way of highlighting the relative sizes of the three age groups is to use **triangular graphs**. The information for the example in Figure 1.18 includes a number of countries within the LEDC, MEDC and **newly industrialised country (NIC)** categories, and the smaller areas within the clustering lines that have been added to the graph show how similar the population characteristics tend to be in each category.

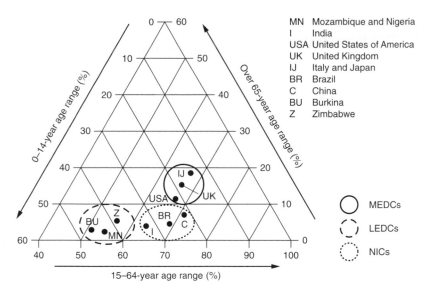

Figure 1.18 Triangular graph to show selected dependency ratios

Summary

- Natural population change is the difference between the birth rate and death rate.
- The world's population grew slowly until the mid-twentieth century, when a 'population explosion' took place due to decreasing death rates in LEDCs.
- Some countries (e.g. China) have adopted anti-natalist population policies to limit their population growth; a small but increasing number of countries (e.g. Sweden) are now adopting pro-natalist policies due to their negative rates of population change.
- The Demographic Transition Model shows changing birth, death and population change rates; a fifth stage has been added recently to include countries experiencing population decrease.
- Malthus believed that the world's population growth would exceed any increases in food and other essential resources, i.e. grow beyond its carrying capacity; his pessimistic view has since been supported by Ehrlich and the Club of Rome; much more optimistic views have been taken by Boserup, Durkheim, Lomborg and Simon.
- The world's food supply has been greatly increased by a series of agricultural revolutions, culminating in high-tech agribusiness techniques such as chemical applications and irrigation.
- Population totals are obtained by means of a census. The first 10-yearly census in England took place in 1801 (population 9.2 million); the 2001 indicated England's population to be 49.1 million and the UK's to be 58.8

million. The greatest inter-census regional population increases had taken place in the south of England and the lowest in the north of England; the only UK country to loose population was Scotland.

*   The composition of population structures can be shown as population/age–sex pyramids; these are based on five-year age ranges called quinquennials and are usually subdivided into childhood (0–14 years), adulthood (15–64 years) and retirement phases.

Student Activities

Research projects

1. Undertake a piece of research similar to that of the case study based on Leyland's enumeration districts – but for an urban area in your own locality. Because some examination specifications require knowledge of how census data changes over a period of time, this project should take the form of a comparison of enumeration district data *changes* during an inter-census interval of at least 20 years. Initial census data research could be directed through your local authority's planning department. The project's conclusions should be in two forms: **a)** a list of significant comparisons between the two sets of census data; and **b)** suggested explanations for the most significant of these comparisons.

2. Undertake a mortality survey for the years 1800–2000 similar to that in the case study based on St Mary's Churchyard, Barton but in an *urban* churchyard environment. Use your survey findings to construct an equivalent table and the two-decade mortality line graphs based on it. This will enable you to create a series of bullet-point observations about the research data. Your final tasks will be to make a series of comparisons between your own bullet-point observations and those given at the end of the case study, then suggest reasons for the most significant differences highlighted by your comparisons.

3. China is one of the world's most rapidly developing countries. Using the Internet and or other resources available to you, undertake research to discover: **a)** reasons why China's economy is expanding so rapidly at the present time; and **b)** the impact of its expanding economy on key factors such as the rate of urbanisation and the many aspects of the country's changing 'quality of life'.

Examination-style questions

1. **a)** With regard to population and resources, explain what is meant by the 'optimistic' and 'pessimistic' views of people such as Boserup and Malthus.

    b) Which of the above two views appears to be the more convincing? Give detailed reasons in support of your decision.

2. **a)** Define each of the following key terms: birth rate, death rate, infant mortality rate, rate of natural increase.

b) Explain how the Demographic Transition Model may be used as a means of predicting future population trends and service provision requirements.
3. Study the display of population pyramids in Figure 1.16 and then answer the following questions:
   a) In what ways did the composition of Britain's population change over the period 1871–2001?
   b) In what ways are the population structure characteristics of typical LEDCs and MEDCs different from each other?
   c) Suggest reasons for each of the differences you have identified in **b)**.
4. With reference to named countries, describe the different types of strategies used in population policies intended to influence national growth rates.
5. **a)** Outline the functions of a modern demographic census.
   b) Summarise the contrasting factual data that has been obtained as a result of censuses undertaken by named LEDC and MEDC countries.

2 Factors Influencing Mortality Rates

Acquired immune deficiency syndrome (AIDS) consequences of HIV infection, including infections, cancers and, ultimately, premature death

Contagious disease disease capable of being passed on by direct physical contact with a diseased person's body or clothing

Deprivation cycle self-perpetuating poverty state from which it is very difficult to break free

Diet the quantity and range of food consumed

Endemic disease disease affecting a population over many generations

Ethnic cleansing use of fear and intimidation by ethnic/religious groups that forces other such groups to migrate against their wishes

Genocide an act that deliberately inflicts conditions of life calculated to bring about a community's destruction in whole or in part

Gross domestic product (GDP) total value of a country's economic output during a year, *excluding* 'net' income gained from overseas trading (the difference between income from overseas investments and the income earned by foreign investment in the domestic economy)

Human immunodeficiency virus (HIV) a virus that destroys the body's protective immune systems, leading to infections and cancers in the later AIDS stage

Infant mortality rate proportion of infants dying between birth and the age of 12 months per 1000 births

Life expectancy average number of years from birth that people in a community may expect to live

Longevity increase in life expectancy due to improved quality of life

Malnutrition lack of a balanced diet due to inadequate intake of protein, carbohydrates, vitamins and minerals

Obesity bodily state of being significantly overweight

Pandemic epidemic existing on an international scale

Quality of life average of three key human characteristics – literacy, life expectancy and infant mortality – by which standards of living in different countries may be compared statistically

Transmittable disease a disease that may be passed from one person to another by airborne, faecal-borne or direct physical contact

Under-nourishment lack of a sufficient *quantity* of food

World Health Organization (WHO) UN agency established in 1948 to encourage international co-operation towards achieving improvements to global health conditions

1 Introduction: The Final Exit

This chapter should be of considerable interest to everyone who reads this book because it's about **mortality**. It is often said that the only universal fact about life is that it has to end in some way, and the two parts of this second chapter examine some of the world's most important – and more interesting – current mortality trends. Many people simply die of old age when the body has reached an advanced state in which quite minor infections and/or organ deficiencies prove to be terminal. Other people die prematurely, due to the actions of others or possibly as a direct result of their own, unwise chosen way of life.

It is a challenging task to produce a simple way of categorising the numerous causes of human mortality. The following four-category model is only one way of doing this and the reader is most welcome to refine and improve it:

- Category 1: death due to normal (i.e. not *induced*) biological causes, for example:
  – old age
  – disease and infection.

  Category 2: accidental death (i.e. no *pre-meditated* actions by humans are involved), for example:
  – natural hazards such as earthquakes, volcanoes and floods
  – pollution incidents
  – genuine 'accidents' such as road traffic accidents (RTAs), ship-wrecks and aircraft crashes; the first RTA fatality was Bridget Driscoll, killed in London on 17 August 1896, since when 25 million people have died worldwide as a result of RTAs and approximately another five times that number of casualties injured or crippled.

  Category 3: pre-meditated and/or 'human conflict' death, for example:
  – executions and extreme torture; the Spanish Inquisition alone executed over 10,000 alleged heretics under its Inquisitor General, Tomas de Torquemada (1420–98)
  – ethnic cleansing and genocide; the Holocaust undertaken by the Nazi Third Reich during the Second World War eliminated 71.4% of all Jews in the Netherlands, 69% in Hungary, 25% in Germany and 22.1% in France
  – wars (including 'civil wars')
  – riots
  – terrorism and sectarian violence; the Northern Ireland 'Troubles' resulted in more than 3500 deaths from the Catholic and Protestant communities and the British and provincial security forces

- murder; the serial killer Dr Harold Shipman murdered at least 215 of his elderly patients while a GP in northern England from the 1970s to the 1990s.

Category 4: self-induced death, for example:
- suicide
- euthanasia.

2 Diseases and their Transmission

This section provides some basic facts about those diseases that have had the greatest impact on global mortality, how they are transmitted between people and what measures can be taken to combat them (Figure 2.1). Figure 2.2 illustrates the cycles by which some of these diseases are transmitted.

The battle against disease is constant and seemingly never-ending because, as humans develop bodily resistance against them, many diseases produce new strains that are increasingly difficult to overcome. One of the worst examples of an **epidemic** on a pandemic scale occurred immediately after the First World War, when about 50 million people already significantly weakened by four years of conflict succumbed to a virulent form of influenza. The AIDS and SARS pandemics of recent times are examined in detail later in this chapter.

Transmitting agents	Diseases transmitted by these agents
Bacteria	Cholera
	Diptheria
	E. coli 0157
	Plague
	Pneumonia
	Salmanella
	Tuberculosis
	Whooping cough
Parasites	Bilharzia
	Malaria
	Sleeping sickness
Viruses	Dengue fever
	Influenza
	Rabies
	Yellow fever

Figure 2.1 Transmitting agents for selected diseases

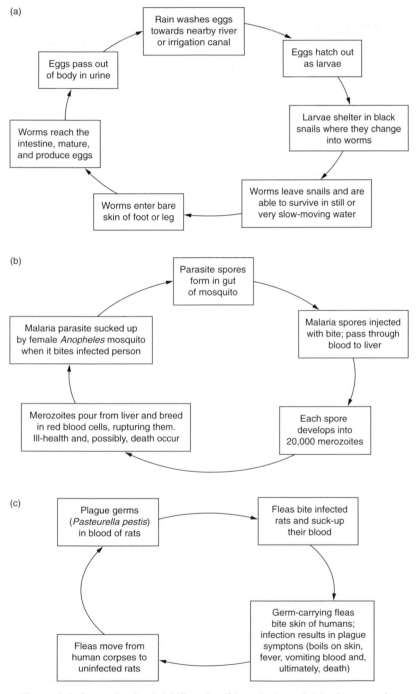

(a)

Rain washes eggs towards nearby river or irrigation canal

Eggs pass out of body in urine

Eggs hatch out as larvae

Worms reach the intestine, mature, and produce eggs

Larvae shelter in black snails where they change into worms

Worms enter bare skin of foot or leg

Worms leave snails and are able to survive in still or very slow-moving water

(b)

Parasite spores form in gut of mosquito

Malaria parasite sucked up by female *Anopheles* mosquito when it bites infected person

Malaria spores injected with bite; pass through blood to liver

Merozoites pour from liver and breed in red blood cells, rupturing them. Ill-health and, possibly, death occur

Each spore develops into 20,000 merozoites

(c)

Plague germs (*Pasteurella pestis*) in blood of rats

Fleas bite infected rats and suck-up their blood

Fleas move from human corpses to uninfected rats

Germ-carrying fleas bite skin of humans; infection results in plague symptons (boils on skin, fever, vomiting blood and, ultimately, death)

Figure 2.2 Stages in the (a) bilharzia, (b) malaria and (c) plague cycles

CASE STUDY: THE SARS PANDEMIC

The first cases of what became known as **severe acute respiratory syndrome (SARS)** were reported in March 2003 in the Guangdong province of China, when people were starting to display unusually severe symptoms of pneumonia. At that time, scientists were unsure about what was causing this debilitating and potentially lethal condition, as well as how it could be transmitted between people. Panic set in shortly after six guests at the Metropole Hotel in Hong Kong caught the infection from a Chinese doctor who was staying there. Inevitably, the worst panic occurred in the regions where the numbers of infected people seemed to be increasing most sharply. Beijing's famous and usually very crowded Tiananmen Square became as deserted as it had been immediately after the massacre in 1989. When SARS began to spread globally, entertainment, tourism, non-essential travel and shopping for luxury items virtually ceased throughout south-east Asia. It had become common knowledge how easily SARS could be transmitted between people who had any form of close contact. Such contacts occur frequently, for example in the home, during medical/dental examinations and when many people are in congested, confined spaces such as aircraft and commuter trains.

SARS usually begins with a high fever (with a body temperature of over 38°C). Later symptoms include head and muscular aches, sudden loss of appetite, dry cough and very unpleasant shortness of breath. The most severe sufferers have to be placed on a breathing respirator within five days of infection. Between February and July 2003, when Taiwan was the last country to be removed from the **World Health Organization (WHO)** list of infected areas, 8437 cases of SARS infection had been reported and at least 819 deaths attributed to the disease – a fatality rate of over 10%. It is possible that both the infection and fatality figures are somewhat lower than reality because the Chinese government initially instructed hospitals *not* to report cases. It was only on 2 April 2003 that China finally gave permission for a WHO medical inspection teams to visit Guangdong, still its most affected area. On 16 April WHO confirmed that SARS was caused by a new virus, a coronavarus, previously not found in humans. To date, a guaranteed cure has not been discovered, although steroids and the **anti-retroviral drug** ribavrin given to patients in Hong Kong have so far proved effective if taken early enough.

Figure 2.3 shows national statistics for SARS-related infections and fatalities during the four-month period of the pandemic. The global cost of SARS has been calculated by the Australian National University at over £22 billion. The World Bank estimated that the national economies of Asia suffered a 1% decrease in GDP. Airlines' revenues were cut by £1 billion and shops in Hong Kong lost £1.2 billion in reduced sales. Although SARS resulted in only a tiny fraction of the number of fatalities due to the post-First World War flu pandemic, it did paralyse many normal global activities throughout the short period of its existence.

Country	Total number of cases	Number of fatalities
Australia	5	0
Brazil	1	0
Canada	250	44
China	5327	348
(Hong Kong)	1755	298
Colombia	1	0
Finland	1	0
France	7	1
Germany	10	0
India	3	0
Indonesia	2	0
Ireland	1	0
Italy	4	0
Korea	3	0
Kuwait	1	0
Macao	1	0
Malaysia	5	2
Mongolia	9	0
New Zealand	1	0
Philippines	14	2
Romania	1	0
Russia	1	0
Singapore	206	32
South Africa	1	1
Spain	1	0
Sweden	3	0
Switzerland	1	0
Taiwan	671	84
Thailand	9	2
UK	4	0
USA	75	0
Vietnam	63	5
GLOBAL TOTALS	8437	819

Figure 2.3 Countries affected by the SARS pandemic

3 The AIDS Pandemic

The following facts give some indication of the scale of the **acquired immune deficiency syndrome (AIDS)** pandemic both globally and within the continent of Africa.

a) Global facts about AIDS

*   Approximately 40 million adults and three million children currently suffer from AIDS worldwide.
*   An average of 6000 people die each day as a result of AIDS.
*   In 2001, 12 countries in sub-Saharan Africa accounted for 70% of all AIDS orphans across the globe.
*   Between 2001 and 2010, the percentage of children orphaned by AIDS as a proportion of all orphans is predicted to rise from 12.4 to 24%.
*   At least 20 million people have died of AIDS since it was first identified in 1981.
*   Ninety per cent of **human immunodeficiency virus (HIV)** cases in children are due to their mothers transmitting the disease during childbirth – a risk that can be virtually eliminated by education programmes and medical support.
*   The price of 'branded' AIDS drugs is often more than twice as high as the price of 'generic' copies produced by companies that are in direct competition with those whose research and development departments created the original, branded drugs. In recent years, such pharmaceutical companies have been forced to reduce the prices of their products.

b) Facts about AIDS in Africa

*   Eleven million children have already been orphaned in sub-Saharan Africa due to AIDS.
*   Over 24.5 million people in sub-Saharan Africa are believed to be HIV-positive.
*   Sixty per cent of 10–14 year olds will be orphaned in Botswana by 2015 because of AIDS.
*   AIDS killed more than 12,000 teachers in Botswana during 2002.
*   By 2005, $10 billion will be needed *every year* to fight the AIDS pandemic within Africa alone; less than half of that sum is now spent in all low- and middle-income countries added together.

c) The detrimental effects of HIV/AIDS

Some of the most serious of these effects are:

*   Many children become orphans and are left in the care of grandparents because their both parents have died due to **mutual**

contamination. Grandparents are often the least well-placed people to care for young children because they may be feeble, under-nourished and cannot access the state or private pensions available in MEDCs. Also, their traditional means of support in old age – their own sons and daughters – has been denied them because of premature AIDS death.

- Effective HIV/AIDS drugs are expensive – partly because of the costly research investment needed to develop them, but also because they are ineffective unless used in long-term, multi-drug cocktail doses and few LEDCs can finance effective drug programmes.
- The people most directly affected by AIDS tend to be sexually active young adults. These are the very people that have the best potential to create wealth and manufacture goods that can then be exported to earn the foreign currency needed to buy drugs in international markets.
- LEDC agriculture tends to be labour-intensive and large reductions in the manual workforce can seriously affect food production. AIDS sufferers are so physically weakened that they are unable to undertake physical work and LEDC countries cannot afford to replace a reduced workforce with large, modern units of agricultural machinery; such machines are designed for extensive farming in MEDCs and cannot function effectively on small and often **fragmented** (widely separated) parcels of land.
- Scarce hospital facilities are already too stretched coping with patients suffering from endemic diseases such as malaria, bilharzia and river blindness, and cannot meet the additional financial and staffing burdens put on them by increasing numbers of AIDS sufferers.
- HIV is frequently transmitted from mother to baby via breast feeding because a baby's underdeveloped stomach allows the virus to enter the bloodstream much more easily. Additionally, it takes 18 months for babies to produce their own protective antibodies, which means that HIV is a major contributor to increasing infant mortality rates.
- HIV/AIDS weakens a person's ability to earn a living, forcing many individuals to adopt prostitution as a means of survival, even though this type of behaviour is known to carry a very high risk of infection to both themselves and their clients. Its practitioners are only too well aware of the risk of infection to others. A woman in Cyprus who had deliberately infected a number of male partners was convicted of pre-meditated murder.
- One alternative to prostitution is the theft of property: looting, burglary and mugging. These activities undermine society as a whole as well as individuals' quality of life. They also place additional pressure on police forces whose strengths may already have been reduced by AIDS – partly as a direct result of physical contact with infected criminals.

CASE STUDY: AIDS IN SOUTHERN AFRICA

a) AIDS in Botswana

The rate of infection fuelling the spread of AIDS in Africa is so severe that some countries face possible 'extinction' if the efforts to control the spread of HIV do not prove more effective. Diamond-rich Botswana is a primary example of this. Although its per capita **gross domestic product (GDP)** is seven times greater than the average for sub-Saharan Africa, its life expectancy has dipped below 40 years for the first time since 1950. Among the country's 1.6 million people, 39% of adults are infected with HIV, with rates over 50% in the north-east and among expectant mothers in cities. Botswana's HIV prevalence rate (35.4%) in 15–49-year olds is the highest in Africa. Approximately 300,000 people in a population of just under 1.7 million are estimated to be HIV positive. AIDS is actually starting to destroy Botswana's workforce and it is predicted that the country's economy will be reduced by one-third as a result of HIV/AIDS.

Botswana's government has sought to control the disease by focusing on prevention. In 2002 it became the first African country to provide free anti-retroviral (ARV) drugs for its people, with the aim of extending the life expectancy and increasing the productivity of people living with the disease. The 2003 cost of the ARV therapy programme was about $8.1 million according to the government's National Strategic Framework for HIV/AIDS 2003–2009. Although Botswana's diamond mines make it one of the wealthiest countries in Africa, it relies on the assistance of international partners to manage and finance the ARV programme, most notably the African Comprehensive HIV/AIDS Partnership (ACHAP). This public–private partnership between the Bill and Melinda Gates Foundation, the US pharmaceutical company Merck and the Botswana government provides technical and financial support to the HIV/AIDS programme. ACHAP had committed $45 million to HIV/AIDS programmes by the end of 2002.

At the 530-bed Princess Marina Hospital, the occupancy rate is described as 'well over 100%' and AIDS treatment dominates its medical care programmes. In the adult medical ward, 70–80% of cases are AIDS related. Kgalaleo Ntsepi is a typical patient at this hospital. She waited until she was very sick before being tested. Her first course of treatment caused severe side-effects, but she stayed on the programme. She now takes Combivir (AZT and 3TC) and Stocrin – a total of five tablets a day – experiences no side-effects and believes that she is as fit and strong as people who are not infected. The compliance rate among the patients receiving ARV treatment at Princess Marina Hospital is 95%, which proves that even poor and often illiterate people are capable of following a strict regime of medical treatment.

b) AIDS in South Africa

In South Africa AIDS primarily affects women. More women than men carry the virus, they are infected at a younger age and they die earlier. It is the mothers, wives, sisters and daughters who have to give up their jobs and drop out of school to care for dying relatives. Two-thirds of those caring for AIDS patients in their final year of life are female relatives. The country's first national study of HIV prevalence reported that 17.7% of women between the ages of 15 and 49 were HIV positive, compared with 12.8% of men; it also indicated that the highest rate of new infections occurs in women aged 15–20. HIV has been identified as the leading cause of death in pregnant women.

One reason for the higher rate of infection in females is biological. The virus does not survive long outside the body, but it does stay alive in the vagina long enough to be able to enter the bloodstream. Another reason for the higher infection rate in women is due to the **patriarchal** nature of African society. The status of women is low, and rape and domestic violence are very common. Over 40% of rape cases in recent years have involved girls under the age of 17. Rape hugely increases the chances of HIV infection because of the internal injuries it causes.

4 Death Due to Natural Hazards

On 26 December 2004 an earthquake occurred in the Indian Ocean close to the western tip of the Indonesian island of Sumatra. Registering an unprecedented 9.0 on the **Richter Scale**, this earthquake triggered a **tsunami** that devastated parts of Thailand, India and Sri Lanka, as well as the nearby Indonesian coastline. It is believed that the number of fatalities attributable to this tsunami exceeded 250,000. Figure 2.4 shows that this human tragedy was by no means an isolated event, and that some regions of the world are particularly vulnerable to certain types of natural hazard.

5 Death Due to Pollution

Most pollution incidents fall into one of two categories, those that take place over a long period of time and those that take place over a much shorter period of time. An example of a long-term event is the

Hazard type	Date	Location	Number of fatalities
Avalanches			
	12/13/16	Dolomite valleys, Italy	18,000
Cyclones/storm surges/tidal waves			
	1228	Netherlands	100,000
	1900	Galvaston Island, USA	8,000
	1970	Bangladesh	300,000+
	1974	Bangladesh	400,000+
	1991	Bangladesh	139,000+
Jan. and Feb.	1953	North Sea Basin	1,800+
	1998	Papua New Guinea	3,000+
Droughts/famines			
	1907	China	24,000,000
	1941/2	China	3,000,000+
	1965/7	India	1,500,000
	1973	Ethiopia	~300,000
	1991	Somalia	350,000
Earthquakes			
	1201	Egypt and Syria	1,000,000
	Feb. 1556	China	~830,000
	Dec. 1920	China	~235,000
	Jan. 1923	Tokyo, Japan	~150,000
	July 1976	Tangshan, China	~500,000
	Aug. 1999	Golcuk, Turkey	17,000
	Jan. 1995	Kobe, Japan	5,500
	Dec. 2003	Bam, Iran	26,300
Extreme temperatures			
	August 2003	Exceptionally high summer temperatures across W Europe (reaching 48°C in France)	35,000 (10,400 in France alone)
	2000s	Hypothermia amongst elderly in England and Wales	20,000 per year
Floods			
	1931	China	3,700,000
	1999	Venezuala	50,000
Volcanic eruptions/lahars			
	Aug. 1979	Vesuvius, Italy	5,500
	Aug. 1883	Krakatowa, Indonesia	36,000
	Nov. 1985	Nevado del Ruiz, Colombia	22,000
Tsunami			
	Sept. 1923	Tokyo, Japan	~150,000
	Dec. 2004	Indian Ocean	250,000+

Figure 2.4 Deaths associated with selected natural disasters

thousands of tube wells that were bored in Bangladesh to avoid repetitions of the cholera epidemics that ravaged that country at frequent intervals until the 1960s. These have now caused arsenic poisoning in at least 43,000 villages and may have accounted for about 10% of the annual national death rate. Another example is that of Minamata considered in the case study on pages 48–49. Examples of those that took place over a short period are the nuclear explosion at Chernobyl and smog.

CASE STUDY: MORTALITY DUE TO POLLUTION INCIDENTS

a) Chernobyl nuclear reactor failure

On 26 April 1986 a catastrophic explosion took place in Unit 2 of the Chernobyl nuclear power plant in a region of the former Soviet Union that is now Ukraine. The primary cause of this explosion was a runaway chain reaction and subsequent fire in the reactor core. This blew off the protective roof of the reactor building and ejected a fine radioactive dust so high that it could be transported great distances by air currents in the atmosphere (see Figure 2.5). Amazingly, the disaster sequence occurred after an emergency cooling system had been deliberately turned off to let an *unauthorised* safety experiment take place.

The highly **radioactive dust plume** that entered the atmosphere had two main components. Iodene-131 was mainly responsible for triggering cancer of the thyroid gland of local inhabitants,

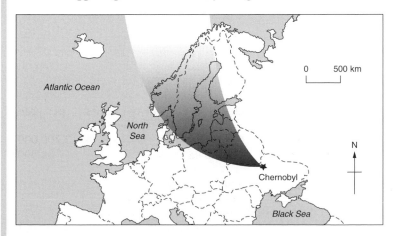

Figure 2.5 Location of the Chernobyl nuclear power station in Russia. The shading shows the area of greatest radioactive fall-out immediately after the explosion, and the intensity of the shading indicates radiation concentration.

particularly in children. Caesium-37, a much longer-living iso-
tope whose radioactivity decreases naturally by only half every 30
years, was the chief cause of whole-body radiation exposure over
much of Europe and, to a lesser extent, throughout the northern
hemisphere. Many of the plant workers and the site restoration
teams, known locally as 'liquidators', received particularly high
doses of radiation. The official figures put the number of fatalities at only 45, but Greenpeace Ukraine still maintains that the
total death toll is likely to have exceeded 32,000. According to
UN estimates, 375,000 Russians were forced to leave their homes,
the majority of these migrants never to return.

b) Minamata Disease

One of the most serious examples of deaths due to marine
pollution occurred during after the Second World War in the
industrialised fishing town of Minamata (population of 35,000)
on the most southerly Japanese island of Kyushu (see Figure 2.6).
The town's name has now become enshrined in environmental
history due to the so-called **Minamata Disease** that ravaged its

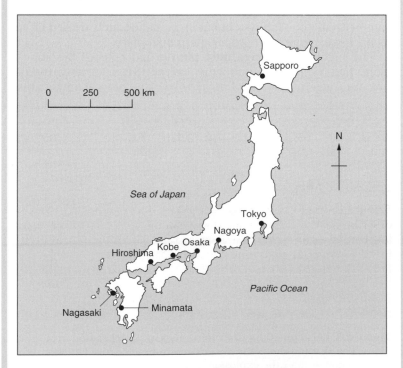

Figure 2.6 Location of Minamata, Japan

population and peaked in the early 1970s. It was caused by the discharge of industrial mercury waste into the local fishing grounds within a sheltered bay to the west of the town. The dangers of inhaling mercury vapour have been known for many centuries. There are well-documented cases of 'madness' among nineteenth century hat-makers, who constantly used mercury compounds in the treatment of the fabrics; hence the Mad Hatter's extremely odd behaviour in *Alice in Wonderland*. The case in Minamata was somewhat different, because it involved inorganic mercury compounds which, being easily converted into soluble forms, can enter food chains and then affect cell membranes within the host's body.

In the 1930s, the Chisso Corporation's factory in Minamata began to produce **acetaldehyde** – a substance involved in the manufacture of a wide range of products including medicines and perfumes. The waste from their new process formed a mercury sludge, which was pumped into Minamata Bay. As the factory expanded, so did the volume of this effluent. After about 20 years of continuous waste dumping, the inhabitants of Minamata began to notice some very unusual and disturbing sights. Large numbers of fish appeared to be in great distress; flocks of sea birds became so weak that even attempting to fly exhausted them; domestic cats began to dance frantically on the streets; by the late 1950s, very few cats were to be seen in the whole of Minamata. At first, the local people were very curious at what they could see, but their feelings changed quite suddenly when some of the human population began to display symptoms of illness. Medical examinations proved beyond doubt that mercury waste was the cause of the paralysis and mental illnesses of the patients. Minamata Disease resulted in over 2250 fatalities and seriously ill patients, 6% of the town's total population.

c) London smog

Smog (a combination of *smoke* and *fog*) is a general term used to describe heavily polluted air. It was commonplace in British cities for centuries and the first laws against polluting the air over London were passed as long ago as 1273. In the reign of Edward I (1272–1307) some Londoners were tortured and executed for breaking these laws, but even such drastic measures failed to deter people from polluting the air around them. Industrial development and the widespread use of coal by industry and for domestic heating increased so much that the number of smoke-haze days in London tripled during the nineteenth century.

The worst recorded incidence of smog pollution in London occurred in late 1952, when a series of climatic factors combined to produce five consecutive days of smog which was so dense that people had to wear protective face-masks. By the end of the fifth day, the city's hospitals could no longer cope with their intakes of critically ill patients. The number of deaths during and immediately after the smog exceeded 4000. As a result, the government passed **Clean Air Acts** in 1956 and 1968, which established **smokeless zones** in all major urban areas. Further legislation demanded even tighter controls on sulphur and lead emissions and much higher **stacks** (power-station chimneys) were constructed for electricity generating stations to reduce local air pollution levels. Figure 2.7 shows the chemical processes involved in smog formation.

$$S + O_2 = SO_2$$

S		O_2		SO_2
Sulphur	+	Oxygen	=	Sulphur dioxide

(This is the standard 'fossil fuel combustion' formula)

SO_2		H_2O		H^+		HSO_2^-
Sulphur dioxide	+	Water droplets in air	=	Hydrogen ion	+	Bisulphate ion

(Sulphur dioxide dissolves readily in damp atmospheric conditions)

$$2HSO_3^- + O_2 \rightarrow 2H^+ + 2SO_4^{2-}$$

(Traces of metal contaminants act as catalysts in the conversion of dissolved sulphur dioxide to sulphuric acid. As sulphuric acid has a great affinity for water, the acidic droplets in the air absorb more water, increasing in size as they do so and making the smog progressively more dense)

Figure 2.7 Chemical processes in smog formation

6 Genocide and Ethnic Cleansing

The word **genocide** was first used in 1946 by a Polish lawyer called Raphael Lemkin, a Jew who lost all but four of his family of over 50 members during the Holocaust. The term **ethnic cleansing** became commonplace during the Yugoslav Conflict of the 1990s. Although similar in some ways, because the overall aim of both is to produce a **homogeneous** nation-state with a single, universally accepted culture, the tactics involved in achieving this state of affairs can vary considerably. Genocide is pure mass-murder. Ethnic cleansing invariably involves the assassination of prominent citizens and systematic murder of ordinary people, but an additional dimension is the widespread out-migration of **refugees** and **asylum-seekers** due to the intolerable levels of intimidation to which they have been subjected. For example, 2.5 million people were displaced as a result of warfare and prolonged harassment during the conflict in former Yugoslavia. The two case studies that follow provide examples of both categories of what are regarded by their perpetrators as justifiable ways of achieving 'national purification'.

CASE STUDY: ETHNIC CLEANSING IN SUDAN

The children of rape (**The Sunday Telegraph,** *25 July 2004*)
In the desperate, sprawling and rain-lashed refugee camps of Darfur, a new breed of outcast is about to be born: children of rape.

When the Janjaweed Arab militiamen rode into Fatima's village of Delibaya eight months ago, she sought refuge with her neighbours in the mosque, hoping that she would be safe there from her fellow Muslims. She was wrong! The attackers were intent on a campaign of ethnic cleansing against black villagers from western Sudan. After they had finished murdering the men, rounding up the goats, cows and camels and burning the homes, they turned their attention to the women sheltering in the mosque.

Fayima was brutally raped and will give birth any day now. Her only shelter, a pitiful mud-floored grass hut, soaked by the onset of torrential downpours in recent days, is a miserable and dangerous enough place to bring a baby into the world. But she has a terrible added burden: the knowledge that even if her baby survives the filthy disease-prone Zalinhei refugee camp in western Darfur, it will be rejected by many in her tribe as the child of their enemy. She can expect to be ostracised by her own people, although she has really not done anything wrong.

Fatima, who is in her late thirties, can barely bring herself to talk about her plight. She stares impassively at the ground, cowed by her suffering. Her words are barely audible. 'My main problem now is being pregnant by the Janjaweed', she mutters. 'My community think this is an illegal child. They have a fight with the Janjaweed so they will not accept the child.' She is already raising her first child alone as her husband has disappeared, presumably murdered in a Janjaweed rampage. They walked on foot for eight hours to reach the camp after the attack and are surviving on handouts – just two small meals a day.

Another victim, Salma, who lives in the Zalingei camp with her two young sons, will soon give birth after being repeatedly raped by three Janjaweed militiamen when she left her village to harvest the family's crops in November. The women in her village washed her in hot water when she returned, a traditional local treatment for victims of sexual attack. Her husband, however, has already rejected her and ignored her pleas for money. 'When I discovered I was pregnant, I just felt very miserable and sad,' she says. 'This is an illegal child and belongs to a man I don't know. I told my husband in Khartoum but he refused to come. I sent a message and asked him to send me some money, but he hasn't replied.' Tears well into her eyes, but she remains defiant. 'When I have the baby I will treat him like my other children. The tribe will not accept the child, but I have decided that he is going to be one of my own.'

The sexual assaults continue, especially as women must still venture out from the safety of the camp to collect firewood from a nearby valley where a group of Janjaweed are based. There are few men in the camp, and for them to make the same journey means certain death. Three weeks ago, Khadia was collecting wood when she was attacked by militiamen who tried to mutilate her sexually with a knife. She fought them off but only by grabbing the blade, receiving deep and still unhealed wounds to her thumb and two fingers. The women recounted their harrowing stories to Fiona Callister, an aid worker with the British charity CAFOD, who has just returned from Darfur. Now that the rains have started, aid agencies are desperately trying to build enough latrines to prevent epidemics of water-borne diseases such as cholera, diarrhoea and typhoid. Plastic sheeting and blankets are being delivered as even those refugees lucky enough to find any shelter are living in pathetically exposed huts and lean-tos (see Figure 2.8).

The rains are seriously hampering food deliveries as usually dry **wadis** (river valleys) turn into impassable quagmires. Some camps have already been cut off and will be reliant on aid drops by helicopter. 'These people are totally dependent on aid to survive,' said Paul Hetherington of Save the Children in Khartoum.' 'Their whole way of life has been destroyed and it hasn't been possible to pre-position enough food in the camps before the rains. It is a very dire situation.'

According to the United Nations, the conflict has already claimed between 30,000 and 50,000 lives and forced 1.2 million people to migrate to nearby countries such as Chad. Fighting erupted nearly 18 months ago when Darfur's black African, mainly Muslim, tribes rebelled. The Arab-dominated regime in Khartoum responded by unleashing the Janjaweed militia, backing their attacks with air raids. Clashes persist despite a recent ceasefire agreement.

The UN said yesterday that two rebel groups from Darfur have agreed to fresh peace talks with the government, a week after the

Figure 2.8 A refugee family in Darfur

first round of talks broke down. The Sudanese government has reacted furiously to a resolution by the US Congress declaring the atrocities in Darfur to be genocide and calling on the Bush administration to intervene. Tony Blair has indicated that Britain might be willing to send troops to the region to protect aid workers. At the UN, diplomats are discussing a US proposal that Sudan arrest militia leaders within a month or face unspecified sanctions. Washington has previously been wary of taking too hard a line against the Khartoum regime, because it is providing important intelligence about terrorist attacks being planned by al-Qaeda!

CASE STUDY: GENOCIDE? THE IRISH POTATO FAMINE

During the 1500s, Ireland was ravaged by almost constant warfare between its English rulers and its Irish inhabitants. As a result of this conflict, Ireland's peasant farmers had great difficulty growing enough food. When the potato was introduced into Ireland in about 1600, it quickly became very popular due to its ability to produce a much higher yield than any crops the Irish had grown before. Also, it could not be easily destroyed by invading soldiers because its nutritious tubers remained underground until it was safe to dig them up.

Ireland had experienced a particularly damp summer in 1845 – ideal conditions for a form of **potato blight** called *Pytophthora infestans* to thrive. This fungus was the principal cause of what has become popularly known as known as the **Irish Potato Famine**. Potato blight reproduces by means of wind- and water-borne spores.

The blight destroyed the Irish potato crop harvests of 1845, 1846 and 1848, as well as those on mainland Europe, causing food prices to soar across the whole continent. Domestic stores of potatoes rotted in cellars and anyone eating them became sick. People were left with very little to eat and had no surplus produce to sell to pay their rent. Many destitute people roamed the countryside, begging for food; some of them ate weeds and grass in order to survive.

Estimates of mortality due to the Irish Potato Famine range from 500,000 to well over 1,500,000, with almost one million being the most realistic figure. The famine also contributed to a significant decrease in the birth rate. Mortality rates were particularly high in parts of western Ireland, due to a serious outbreak of cholera that occurred there in 1849. The most vulnerable sectors of the population were children under five years of age, the old, and pregnant and lactating mothers. The poorest people were especially vulnerable, because their quality of life had already been very low well before the start of the famine. Nearly half of all rural families had been living in windowless, one-room mud cabins in which their pigs also slept. The long-term unemployed slept in ditches covered over by flimsy shelters. Ireland was certainly overpopulated before the famine, with many of its rural families surviving on only half an acre of land or less, relying on their potato crop to feed them during the winter months. Even those who could grow grain or barley were faced with a stark choice: sell their crops in order to pay the rent, or eat food and be evicted by their landlords!

Ireland's 1841 population of 8,175,124 decreased to 6,552,385 in 1851 – almost 37% below the 9,018,799 that would have been likely had the famine not reduced the normal rate of population increase for mid-nineteenth century Ireland.

This case study of Ireland's greatest famine would not be complete without some account of the million people who emigrated as a direct result of it. Curiously, only about 50,000 families were evicted from their homes during the famine for the following reasons. In January 1847 the government announced that in future all destitute people were to be housed and fed at the expense of property owners. The only way the landlords could avoid bankruptcy due to this legislation was to reduce the number of destitute on their estates as quickly as possible by using emigration as an alternative to eviction. The cost of arranging for a pauper family to emigrate was about half the annual cost of maintaining it in the local workhouse. Predictably, the landlords chose the cheapest available means of overseas transport and so most of the poorest emigrants sailed the shortest possible distance to the west coast of Britain. As a result, the Irish population there rose quickly to 13% in Manchester, 18% in Glasgow and 25% in Liverpool; in direct contrast, London's Irish component reached only 5%. The 1841 census listed 289,404 Irish-born people in Britain; by 1861, there were over 600,000. The slightly better-off families emigrated to North America, where they again encountered the potato blight that had already brought about their ruin back in Ireland.

With a few notable exceptions, the rich Irish felt that their only obligation to the poor was to make a donation to charity. Many of them were **absentee landlords**, who lived permanently in England for reasons of personal safety and were almost totally ignorant of the dire conditions under which their Irish workers existed. However, this very same class of landlords continued to make valuable profits through the export of foodstuffs such as grain as well as wool and flax (the crop from which linen is manufactured). Large quantities of wheat, oats, barley, eggs, butter beef and pork were also produced throughout the most traumatic months of the famine, much of it exported to Europe at a hefty profit. It is reported that for every ship bringing desperately needed food to Ireland, six left, fully laden with far more nutritious cargos! The scarcity of food in Ireland was blamed primarily on the weather, the potato fungus, but, certainly the most curious of all, on the Malthusian concept of overpopulation and its 'inevitable' consequence of population reduction.

Could this disastrous famine have been averted, or was it a premeditated form of genocide (defined by the United Nations Convention of Genocide in 1948 as any 'act which deliberately inflicts conditions of life calculated to bring about a group's destruction in whole or in part')? The historian Peter Gray believes that the British government's policy was not so much a deliberate policy of genocide but a stubborn, dogmatic refusal to admit that its policies were inappropriate and that they would inevitably result in the deaths of hundreds of thousands of people. The Irish-American historian Dennis Clark disagrees, stating that 'The British government's insistence on the absolute right of landowners to evict farmers and their families so that they could raise cattle and sheep was a process as close to ethnic cleansing as any Balkan war ever enacted'. Perhaps significantly, the British spent only £7 million on humanitarian relief in Ireland between 1845 and 1850, far short of the £20 million compensation given to West Indian slave-owners in the 1830s following the abolition of slavery. Possibly the most telling 'evidence' of genocide was the British decision to ship large quantities of Indian-grown wheat to Ireland, a type of wheat which was *known at that time* to need milling *twice* to make it safe to eat. The Irish poor were totally unaware of the dangers of eating this imported, single-milled wheat, and many of them died after suffering agonising stomach pain.

7 Unhealthy Life-styles

People can increase their risk of dying prematurely by engaging in life-styles that are unhealthy and therefore potentially dangerous. Primitive humans were able to survive because both their life-style and bodies had adapted to hazardous environments in which the unpredictable was commonplace. Bodily fitness was quickly achieved because even the most basic tasks demanded physical activity. Early peoples' diets were varied and healthy because they included a wide range of nutritious foods. Because it was difficult to keep food fresh for any length of time their bodies evolved to store any excess food as fat, in much the same way as a camel's humps provide essential reserves in hot desert areas. Their bodies were constantly exposed to infection, and so acquired a much higher degree of natural resistance than exists in people today. Some measure of protection was obtained from natural substances discovered by trial and error to be effective and safe to use, as in the case of the Australian Aborigines and the native tribes of the Amazon Basin, who have become expert users of plants which can ease the symptoms of headaches and fevers. Our increasing use of such 'natural' remedies today has its roots in the acquired wisdom of our ancestors. No doubt some overuse of such powerful natural substances did take place, the forerunner of our current issues with drugs, alcohol and smoking. However, early peoples' environments were unpolluted, except by animal life within the local ecosystem.

This section focuses on the problems association with obesity, but it should not be forgotten that smoking, the consumption of alcohol and the taking of addictive drugs are other current issues causing concern. This high degree of concern is not only because these self-pollutants endanger the health of many more people, but also because individuals are themselves becoming increasingly incapacitated by more extreme forms of the condition.

a) Obesity

Obesity is a general term used to describe being excessively overweight. The condition has now been accurately defined, which makes it possible for statistics to be collated using widely accepted units of reference. The illustration in Figure 2.9 represents an advance on earlier tables, which tended to show people as being simply 'underweight', of 'average weight' and 'obese'. From this table, it is clear that any **body mass index (BMI)** between 25 and 30 classifies a person as being merely 'overweight', and that anyone whose BMI is over 30 is within one of the 'obese' categories. Obesity is now so serious and commonplace that the single previous 'obesity' category has had to be subdivided. Doing this assists research analysis and takes account of those people who are actually extending the upper ranges of the obesity scale.

Obesity is a potentially difficult issue, because being overweight can be caused by purely **genetic factors** (factors which are inherited biologically), just as much as overeating, the drinking of sugar-rich beverages and lack of physical activity. Historically, obesity has tended to be associated with overeating, which in turn was linked to constant access to large quantities of rich food; Friar Tuck remains a caricature figure of monks who enjoyed a medieval 'good life'. King Henry III is recorded as saying, 'Let me have men about me who are fat!', because

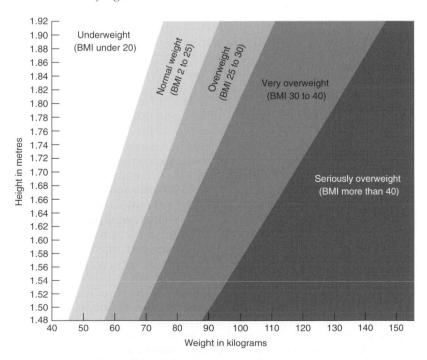

Figure 2.9 Body mass index calculation graph

such people were regarded as content, trustworthy and better company on social occasions. In fact, being overweight can trigger a range of serious medical problems, especially in extreme cases. In the US state of Colorado, one ambulance company had so much difficulty transporting overweight clients that it has now refitted its entire fleet of vehicles with extra-large compartments and winches capable of lifting patients weighing up to half-a-tonne! The next case study provides information about obesity trends within the UK.

It is hardly surprising that many books giving dietary advice are best-sellers, the classic example being the series of books based on the Atkins' Diet, the first of which, *Diet Revolution*, has sold over 20 million copies since 1972 and has been translated into more than 25 languages. Governments are increasingly aware of the medical and economic costs of obesity, and it is probably only a matter of time before legislation puts strict limits on commercial food advertising and the sale of 'unhealthy' items in schools. There is certainly growing evidence that people who regularly eat high dosages of fat and sugar can suffer withdrawal symptoms when their intake is reduced and that such 'addicts' need to eat ever-increasing quantities in order to experience the same levels of satisfaction. In other words, prolonged unhealthy eating can induce cravings that are very similar to those induced by stimulants such as alcohol, smoking, narcotic drugs and even coffee.

It is now scientifically possible to create pills that can reduce obesity problems, but research into such drugs is still in its early stages. Professor Steve Bloom of the Department of Metabolic Medicine at Imperial College, London, believes that what is needed is a drug targeted at controlling a person's *appetite* for food. He has so far identified 114 different drugs that have some capacity to help people to lose weight, but the majority of these induce adverse side-effects and are not particularly effective. He is only one of many people in the medical profession who believe that a much more productive strategy is educating people to eat less and change to more balanced, nutritious diets, especially during young peoples' formative years at school. It is much more difficult (and costly) to change bad eating habits in later life, as thousands of adults have discovered when trying to follow Weight Watchers' slimming programmes. The US government recommends a daily intake of 1600 calories for **sedentary** (physically inactive) women and 2000 for men, but recent surveys undertaken by New York University indicate that the actual calorie intake figures are 1877 and 2618, respectively: a very significant excess in each case. In late 2004, the British government launched a campaign to encourage a reduction in sugar consumption, which was targeted especially at soft drinks and sugar-laden breakfast cereals. It remains to be seen whether this campaign will prove as successful as earlier government initiatives aimed at salt intake reduction, which had a highly beneficial effect on the number of people suffering hypertension, strokes and heart attacks.

In 2002, McDonald's restaurants recorded a deficit of £211 million during its final three months of trading, the first time that the fast-food chain had lost money in its 47-year history. The company has now had to revise its image by offering less fatty products, by using vegetable oil instead of lard in its burger buns and by offering salads as alternatives to traditional favourites such as the 'Big Mac'. McDonald's has also started displaying nutritional information on product packaging. Wimpy have followed suit, with its marketing chief asserting: 'We do have a responsibility to provide a varied menu with healthier alternatives for our customers'. A key factor in this sudden change in the fortunes of many fast-food chains at that time was certainly the publication in the USA in 2001 of Eric Schlosser's book *Fast Food Nation*. This made scathing attacks on the US government, fast-food companies and others associated with agribusiness for their failure to safeguard the health of the US people. Figure 2.10 shows the essential components of a typical balanced, healthy diet. This diagram excludes the vast range of artificial additives contained in modern fast foods. The cumulative effects of these additive cocktails is not yet fully known, although undertakers in Germany have recently discovered that it is no longer possible to reuse burial plots

Dietary component	Bodily function	Food sources
Carbohydrates	Fast-release energy	Cereals, root crops
Fats	Slow-release energy	Dairy products, nuts, red meats
Fibre		Fruit, vegetables
Minerals	Strengthens teeth and bones	Dairy products
Protein	Production and maintenance of body tissues and fluids	Fish, meats, nuts and seeds, tofu
Vitamin A	Healthy skin; good night-vision	Dairy products, fish, red fruit and vegetables
Vitamin B	As for protein	As for protein
Vitamin C	Prevents scurvy	Citrus fruits, vegetables
Vitamin D	Healthy bones and teeth; prevents rickets	Dairy products, eggs, sunshine
Vitamin E	Maintains fertility levels	Oily fish

Figure 2.10 Essential dietary components

in cemeteries as frequently as they used to. Corpses normally decompose within 8–10 years of burial, but the increasing use of food preservatives has retarded the decomposition process so much that plots cannot be re-used for 40–50 years.

b) Smoking

There are 1.1 billion habitual smokers, of whom 82% live in lower- and middle-income countries. However, it is now believed that obesity will soon overtake smoking as the chief cause of cancer. In the USA, where obesity is a much greater problem than in the UK, approximately one-third of all cancers are linked to excessive eating and Cancer Research UK (CRUK) has predicted that a similar proportion of Britons will succumb to cancers within the next 30 years for precisely that reason. Research has highlighted Cambridge as the 'smoking capital' of Britain, with its households buying an average of 25 cigarettes a day; also that Balkan Sobranie is among the best-selling brands – almost certainly the choice of university students and college dons. This phenomenon has since been nicknamed 'The Cambridge Diet', because heavy smoking is a known cause of weight loss! The dangers of both active and passive smoking have been well documented for a long period of time, but it is only recently that governments such as those of the USA, Ireland and Scotland have started to take the initiative in banning smoking in public places.

c) Alcohol

The consumption of excessive quantities of alcohol has caused difficulties for centuries, as it not only leads to serious liver and kidney conditions, but has adverse effects on family life, behaviour patterns and industrial productivity. Of particular concern at the present time in the UK is 'binge-drinking', so called because it involves the deliberate intention of drinking excessive quantities of alcohol within the relatively short period of time available between the opening and closing of licensed premises. This habit has been growing in popularity, especially with women in their late teens and early twenties. Doctors are starting to predict a significant deterioration in female health and a consequent reduction in longevity due to such bouts of excessive drinking.

CASE STUDY: OBESITY TRENDS IN THE UK

The following facts indicate the extent of the current obesity problem in the UK, including some of those trends which have made it one of the nation's issues of greatest concern. The UK has progressed through the **nutritional transition** phase and has now become a well-established '**fast-food**' nation. The health of its people – especially its youngsters – is suffering as a direct result of this.

- In Britain, one-fifth of the population is now clinically obese. Seats at Wimbledon's Centre Court are to be widened by 10%, to 46 cm, to accommodate the expanding waistlines of spectators in time for its 2009 season tennis championships.
- Today's children may prove to be the first UK generation to be outlived by their parents.
- A recent survey in England showed that 18% of school-children are overweight and that a further 6% are within the obese categories. In 2003, the Health Development Agency revealed that 10% of UK six year olds and 18% of 15 year olds were obese – 1.5% and 3% increases, respectively, within a survey period of only two years.
- Some youngsters' health is so severely at risk (Figure 2.11) that doctors now believe they have no option but to start giving children as young as 14 years of age stomach-shrinking surgery, using techniques such as size-limiting staples.
- British children are currently buying much greater amounts of 'sweet things' than they did 50 years ago. Their intake of soft drinks has increased 25 times and confectionery items such as chocolates and sweets 30 times. Many soft drinks contain only

Figure 2.11 Obesity in very young children

8% of natural juices, much of the remaining liquid containing high levels of sugar, colouring, preservatives, artificial sweeteners and flavour enhancers. Seven-to-ten year olds seem to have the greatest inclination to indulge in regular 'snacking' of items that contain sweeteners, artificial colourings and preservatives. On average during 2004, more than 1150 advertisements for '**junk food**' were shown *daily* during and between children's television programmes.

- In the 1940s, the Marks and Spencer standard pattern for adult women's clothing was 33–21–33 inches; today's average British woman measures 36–28–38 inches, which represents increases of 9%, 33% and 15% for the bust, waist and hips, respectively.
- It is believed that about 5% of all new cancer cases in British women (about 6800 cases a year) and 3% in men (about 4000) are attributable to obesity. A recent National Audit Office report estimated that the annual healthcare costs directly attributable to obesity exceed £470 million and that at least 31,000 deaths in the UK were also caused by the condition. The total cost to the national economy is believed to exceed £2.6 billion annually.
- Figure 2.12 shows the percentages of common medical conditions that are directly attributable to obesity.

Disease	Cases attributable to obesity (%)
Gout	47
Type 2 diabetes	47
Hypertension	36
Colon cancer	29
Angina	15
Gallstones	15
Ovarian cancer	13
Osteoarthritis	12
Strokes	6
Prostate cancer	3
Rectal cancer	1

Figure 2.12 Links between obesity and illness

8 Self-induced Mortality: Suicide and Euthanasia

a) Suicide

Suicide and **euthanasia** are both premeditated acts, a fact which sets them apart from the many other causes of mortality discussed at the start of this chapter. Both are extremely sensitive issues, taboo subjects to be avoided whenever possible in both private discussions and public debates. The reasons for such extreme caution are understandable. Most suicides are a direct result of overwhelming personal problems such as:

- drug and alcohol abuse
- divorce and other failed personal relationships
- financial worries, often triggered by unemployment or an inability to manage personal debt
- working excessively long hours
- working with people who are a cause of friction
- the inability to juggle the conflicting demands of home and work.

It is an established fact that women are more likely to attempt suicide than men, but that men are more than four times more likely to actually succeed in ending their lives. Men are notorious for keeping problems to themselves, whereas women are much more likely to discuss personal difficulties with friends and seek medical support. Only 24% of the 5000 adults who kill themselves each year in England have had some form of contact with the medical services. When the depressed Japanese economy in the late 1990s led to widespread company bankruptcies and unemployment, it was the highest-ranking males who paid the ultimate personal sacrifice, shamed into suicide by the failure of organisations that had paid them high salaries to ensure their success. However, fewer than 10% of global suicides take place in MEDCs.

The number of deaths due to suicide is quite surprising:

- about a million suicide deaths occurred globally in 2002. This figure exceeded the total number of deaths due to armed conflict in that year!
- During the past 50 years, global suicide rates have increased by over 60%.
- Suicide is now the single largest cause of death among people within the 15–34 age range – the group in which alcohol, drug and personal-relationship difficulties tend to peak. Figure 2.13 indicates that this is true of both men and women.
- In the USA, the number of suicides is 1.75 times higher than that of **homicides** (murders). This could well be linked to the fact that over 200 million firearms are privately owned by US citizens, and so are readily available for the purpose.

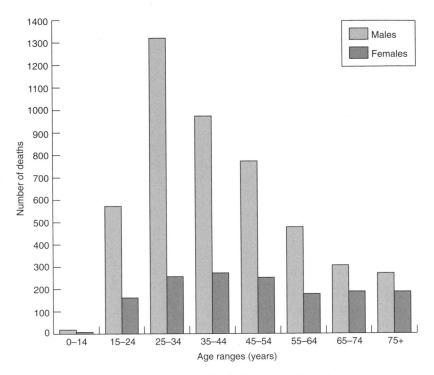

Figure 2.13 Suicide deaths by age and gender

- Some of the world's highest suicide rates occur in the states of the former Soviet Union – especially Kazakhstan, Latvia and Lithuania. Possibly for this reason, the former USSR kept its national records on suicide totally secret until it disintegrated in the 1990s.

It is possible to restrict the opportunities available to people intending to take their own lives. Potential suicide locations such as bridges and cliff-edges can never be rendered completely safe, but measures can be taken to identify and then monitor individuals in suicide-inducing situations. In Britain, the availability of drugs that can be lethal if overdosed is now much more closely regulated – partly to reduce a patient's ability to commit suicide, but also to safeguard doctors who prescribe them against costly litigation.

b) Euthanasia

Suicide is regarded as a highly personal act, achieved without the help of others. Euthanasia has been a 'closed' subject in the past, mainly because it requires the assistance of others, whether these be relatives or trusted members of the medical profession. Both groups of people are well aware that assisting a person to commit suicide is a criminal offence in the UK, although it is common knowledge that *many* doctors privately regard 'mercy killings' as simply one aspect of patient care. There is now increasing support for legislation permitting euthanasia under strictly controlled circumstances that safeguard the interests both of the terminally ill patients and their life-insurers.

In the Netherlands, euthanasia of terminally ill patients carried out in the presence of a doctor was legalised in 2002. Similar measures were approved in Belgium in the same year, but only for persons over the age of 18 who were fully conscious at the time of consent.

CASE STUDY: THE ASSISTED SUICIDE OF REGINALD CREW

Reginald Crew ended his life at the age of 74 in 2003, in an '**assisted suicide**' at a Swiss flat, firmly believing that he would finally achieve the dignity he had been seeking after years of suffering from motor neurone disease. Mr Crew had sought an assisted suicide after gradually losing control of his muscles. By the time he died, he was paralysed from the neck down and incapable of eating solids. He was afraid of losing his power of speech as well. Mr Crew had said that he was afraid not of dying, but of living any longer, and that an assisted suicide was 'the best gift' he could hope for. Mrs Crew had spent the previous four years nursing the former Liverpool dock-worker through the many indignities of the body-wasting disease. She travelled to Zurich with him to meet doctors working for Dignitas, a voluntary organisation that arranges assisted suicides for terminally ill patients from all parts of the world. The Crew family paid no fees as the service is paid for by the Dignitas membership fee of £10 per person. In the past few years, the non-profit making organisation has helped 134 people to end their lives. So far, 14 British people have become members of the organisation. However, many people feel that allowing assisted suicide is a 'slippery slope', which could result in some curable patients dying. The British Medical Association (the doctors' union) is firmly against any such measure, because it could alter the balance of trust between doctor and patient. Also, the authorities in Switzerland are deeply worried about the influx of foreigners to the Zurich-based clinic and are considering passing emergency legislation to prevent this in future.

Summary

- Death may be due to biological causes, accidents and human conflict; it can also be self-induced.
- Epidemics occur when large numbers of deaths are caused by a disease; pandemics are epidemics on a global scale (e.g. SARS and HIV/AIDS). Some diseases (e.g. malaria) may be reduced by interrupting their cycles.
- Natural hazards are a continuing reason for widespread mortality (e.g. the earthquake-induced Indian Ocean tsunami in December 2004).
- Accidents have many causes, including road traffic accidents and pollution incidents such as the Chernobyl nuclear power plant explosion in April 1986.
- Human conflict includes ethnic cleansing (e.g. in Yugoslavia in the 1990s) and genocide (e.g. the Holocaust in Germany during the Second World War); both invariably result in the mass-migration of asylum-seekers and refugees.
- Premature death may be caused by alcoholism, drug-taking, overeating leading to obesity and smoking; it may also be induced by euthanasia and suicide.

Student Activities

Research projects

1. Using newspaper and Internet resources, undertake research to discover the current situation concerning the issues introduced by the following case studies in Chapter 2:

   a) *AIDS in southern Africa*: has the scale of HIV/AIDS infection improved since this case study was written? If so, to what extent has it improved?

   b) *Obesity in the UK*: what recent measures have been taken to reduce this problem? How effective do these measures appear to have been?

   c) *Mortality due to pollution incidents*: Chernobyl and Minamata could be updated or some other major pollution incidents could be investigated (e.g. the world's worst chemical pollution incident at Bhopal, in India, which occurred on the night of 2–3 December 1984).

   d) *Ethnic cleansing in Sudan*: what has been the effect of the widespread international condemnation of recent events in Darfur?

   e) *The assisted suicide of Reginald Crew*: On 30 November 2004, a British judge lifted a travel ban on a woman who had asked her husband to take her to Switzerland so that Dignitas could fulfil her wish of an assisted suicide. In doing this, the judge protected her husband against a prison sentence of up to 14 years for giving such assistance. 'Assisted suicide' is clearly an issue that is going to continue to make headline news for a long time, and so it should be quite easy to monitor its developments since Reginald Crew's death.

2. Use Internet sites to research the tsunami that devastated the Indian Ocean on 26 December 2004 and, in particular, discover:
   a) why the **epicentre** of the earthquake that caused the tsunami occurred very close to the western tip of Sumatra
   b) basic facts about the tsunami, e.g. its wave's height and speed
   c) the numbers of fatalities that occurred along the most affected coasts of the Indian Ocean *and* the chief reasons for this pattern of mortality
   d) what measures were taken to support the affected countries both immediately after the event and in the much longer term.

Examination-style questions

1. Refer to the illustrations on pages 38–39 before answering the following questions:
   a) Name two diseases in each of the three transmitting agent categories: bacteria, parasites and viruses.
   b) With reference to the cycles by which diseases are transmitted suggest ways in which the spread of certain diseases can be controlled.
2. With reference to a wide range of historical events, discuss the validity of the statement: 'That armed conflict has been the chief cause of deaths which are not attributable to failing health in old age'.
3. a) Briefly describe the links between the HIV and AIDS medical conditions.
   b) Summarise the impact of the HIV/AIDS pandemic on populations at a range of scales from national to global.
4. a) State any ways in which 'genocide' and 'ethnic cleansing' may be regarded as both similar and different means of achieving more homogeneous communities.
   b) Give a detailed account of *either* one genocide *or* one ethnic cleansing event that includes the aims which it was intended to achieve as well as the outcomes which resulted from it.

3 Migration

1 Introduction: Types of Migration

Migration is one of the most important aspects of demography and a recent (quite possibly conservative) estimate puts the global number of migrants in a typical year at about 150 million people. If that rate were to be maintained it is theoretically possible for the entire global population of over six billion people to migrate at least once every 40 years!

All migrations are unique, and not just in terms of the numbers of people involved. Every migration can, however, be categorised according to the responses to these key questions:

- *Is this migration likely to be permanent?* The internationally agreed definition of 'migration' is any population movement involving a change of residence of at least one year's duration. It is, therefore, appropriate to use other terms for any shorter population movements. Types of **seasonal migration** include those followed by 'travellers' and fairground workers, as well as **nomads** such as the Bedouin tribes of North Africa whose movements are dictated by the availability of food for their camels and other animals. One increasingly popular form of seasonal migration is the autumn exodus of elderly people escaping from northern winters by flying to milder climate areas further south. **Commuting**, the daily travelling between home and work, is not a true form of migration because it does not involve a change of residence.
- *Was the decision to move taken by the migrants themselves?* **Voluntary migrations** are decided in this way, whereas **forced migrations** involve some degree of coercion. Political factors are the most common reasons for forced migrations, but there are many examples of this that are not due to intimidation. One such common occurrence is the need for people to leave lowland areas that are about to be flooded as part of reservoir construction schemes (e.g. the Aswan High Dam in Egypt which displaced 100,000 people after its completion in 1971 and the Three Gorges Scheme in China which will have displaced 1.2 million by its completion in 2010); another is when natural hazards make it unwise to remain until the volcanic or other threat has receded sufficiently, as described later in this chapter for Tristan da Cunha.
- *What are the chief reasons for this migration?* Most voluntary migrations take place because families are seeking to improve their quality of life. Such people can regarded as **economic migrants**, because obtaining better housing, health and educational provision is often heavily dependent on securing worthwhile, long-term employment. An increasing trend is for older people to move at the end of their working lives (**retirement migration**) to places chosen because they provide a refreshing contrast to those in which they used to be employed.

- *What is the distance involved in this migration, and does it involve crossing international boundaries or areas of sea?* Some of the most challenging of all population movements have been **intercontinental migrations**, across ocean barriers and international boundaries. Migration that takes place completely within a country is referred to as **transmigration**. Examples of short-distance migrations are **suburbanisation** (flows from inner to outer residential zones) and counter-urbanisation (flows from large urban areas to smaller urban settlements and rural areas).
- *What differences between the donor and receiver places have resulted in this migration?* Such differences can take many forms. Rural–urban migration became increasingly common in LEDCs during the second half of the last century due to growing dissatisfaction with village life in the poorer regions and is, therefore, described in considerable detail in this book. In contrast, urban–rural migration has taken place largely in MEDCs and has resulted in the partial depopulation of many cities in such countries. This form of migration flow follows residents' conscious decisions to escape from overcrowding, high property and living costs, social unrest and air and noise pollution within the more densely populated urban areas.

Migration is also noteworthy for the quantity of research that has been undertaken on this topic and the numbers of laws and theories that have resulted from these investigations. The next section examines the most important of these research conclusions, and assesses their relevance to current patterns of migration.

2 Migration Theory: Laws, Models and Principles

The four most influential theoretical models of migration are discussed below in chronological order.

a) Ravenstein's Laws of Migration

The 11 so-called **Laws of Migration** were developed during the second half of the nineteenth century by a **cartographer** (mapmaker) called E.G. Ravenstein whose increasing awareness of changing patterns of population growth and distribution made him curious about the reasons why people migrated between places. His later investigations, based on observations of birthplace data contained in the 1871 and 1881 British censuses, enabled him to examine migration patterns between selected counties during that 10-year inter-census period. His research focused on three key aspects of the migration process:

- the relative importance of rural–urban migrations and **inter-urban migrations**

- the relative mobility of males and females
- the impact of technological developments (especially in transport networks) on the number of people able to migrate between places.

Ravenstein took into account the following two crucial factors:

- the relative size of the migration source and destination centres
- the travelling distance between both of these centres.

Ravenstein published his research conclusions as a series of articles in 1876 in the *Geographical Magazine,* and the *Journal of the Royal Statistical Society* in 1885 and 1889. Of these, his 1885 conclusions have been the most widely quoted. His 'laws' have certainly withstood the test of time and are still generally regarded as being relevant, even though much larger numbers of people are now able to migrate due to the availability of cheaper and quicker forms of mass transportation. The migration theories that followed Ravenstein's own groundbreaking work, which are described later in this chapter, increasingly used mathematical formulae. These theoretical conclusions do, however, tend to reinforce his basic laws and it is for this reason that Ravenstein is now widely regarded as the 'founding father' of modern migration theory.

Ravenstein's 11 Laws of Migration are summarised below, each law being supplemented by a brief explanation, each printed in italics, of its core meaning:

- Law 1: most migrants travel short distances. *This is due to limited technology in transport and communications. Also, people usually know much more about any job opportunities that occur close to their own local area.*
- Law 2: migration usually takes places in **waves**. *This means that migration often takes place in stages, starting with a move from a rural area to a small town, then to a much larger town and finally to a city. Urban migrants rarely return to a rural area, but seem to become 'locked into' the urban hierarchy, with any later migrations taking them to increasingly large urban settlements.*
- Law 3: most migrants travelling long distances intend to go to one of the larger commercial or industrial centres. *People's knowledge of opportunities in distant (especially foreign) places is usually restricted to jobs in the largest cities.*
- Law 4: every wave of migration tends to produce a compensatory **counter-wave**. *For example, the migration of poor people into a town often results in that town's richer people moving out to nearby villages from which they can commute to work in their same place of employment. Such waves are now known as suburbanisation and counter-urbanisation.*
- Law 5: townspeople tend to be less migratory than those living in rural areas. *This is because the greater opportunities in urban areas encourage their populations to remain there.*

- Law 6: many males migrate considerable distances, much further than females, whose travels tend to be within a relatively small radius. *Traditionally, males have been willing to travel long distances to find suitable work, whereas females' migrations have often followed marriage.*
- Law 7: most migrants are adults; entire families rarely migrate out of their country of birth. *This may be explained by the fact that an adult male is usually the first member of a family to migrate. This person is then able to send money back to the other members of the family, until such time as it is physically safe and economically sound for them to re-join the migrant wage-earner in the new home.*
- Law 8: large towns grow more by migration than by natural population increase. *The perceived attractions of large settlements are so great that it is quite usual for the number of migrants to exceed any natural urban population growth (the net difference between birth and death rates).*
- Law 9: migration increases in volume as industries and commerce develop and transport facilities improve. *The number of rural–urban migrants will inevitably rise when more jobs are created due to the expansion of urban areas' **secondary** and **tertiary employment** factors.*
- Law 10: the major direction of migration is from agricultural areas to the centres of industry and commerce. *There are invariably fewer employment, social and recreational opportunities in rural areas.*
- Law 11: the major causes of migration are economic. *Most migrants move for economic reasons, to increase their income and so achieve a higher quality of life for themselves and their families.*

b) Zelinsky's Mobility Transition Model

Mobility transition is a five-stage model devised in 1971 by a US geographer called Wilbur Zelinsky. Its aim was to identify possible links between the nature and intensity of migration patterns and the phases of socio-economic development achieved by countries involved in migration. Its concept is, therefore, very similar to the demographic transition model discussed in Chapter 1, in which economic development is perceived to be a major factor in the changing patterns of birth, death and population growth rates. Each category of Zelinsky's model exhibits different mobility characteristics, as listed below:

- Phase 1: the pre-modern, traditional society:
  - there is little genuine residential migration
  - localised circulation takes place between neighbouring villages purely to meet local communities' agricultural, trading, social, religious and defensive needs.
- Phase 2: the early transitional society:
  - mass migration takes place, involving large numbers of people who move from rural regions to expanding urban settlements within the same country

- overseas colonies attract the rural poor due to the availability of large areas of virgin land suitable for exploitation by pioneer families
- smaller but still significant numbers of skilled workers, technicians and professionals may also migrate between the more developed regions of the world
- there is a significant growth in various kinds of **internal population circulation**.
- Phase 3: the late transitional society:
  - there is a reduction in the rate of migration from countryside to city
  - the flow of migrants to colonisation frontiers is also reduced
  - emigration is declining or may have ceased altogether
  - population circulation continues, often in increasingly complex patterns of movement.
- Phase 4: the advanced society:
  - overall residential mobility tends to level off, but may continue at a higher level between particular regions
  - movement from countryside still occurs, but at a reduced rate
  - significant movements of migrants often occur between cities and within major urban areas
  - **pioneer emigration** may now be stagnant or even reversed in flow
  - there is a significant net immigration of unskilled and semi-skilled workers from less developed countries
  - high volumes of circulation take place, mainly for economic and recreational purposes.
- Phase 5: the future super-advanced society:
  - there may be a decline in the level of residential migration as well as in some forms of circulation due to improved communications technology
  - almost all residential migration may be of the inter-urban and intra-urban types
  - some further immigration of relatively unskilled labour from less developed areas is still possible
  - there are increased volumes of some current forms of circulation; there may also be some new forms of population circulation
  - an increased political control of internal as well as international movements may take place.

The Zelinsky model is generally regarded as a useful (if somewhat over-simplistic) analytical tool. Its critics believe that there are a number of reasons why it should now be considered to be somewhat flawed:

- There is an over-reliance on a vague concept of the ways in which countries develop.

- Its conclusions are based almost exclusively on migration patterns affecting MEDCs.
- It does not take into account cultural factors in migration, which are especially strong in LEDCs.
- It fails to consider counter-urbanisation – one of the most important migration flows of the second half of the twentieth century.
- It under-estimates the extent to which political controls on both international migration and internal movement have been applied.
- It fails to consider reasons why *individuals* migrate; its chief focus is on group movements, at the national and global scales of migration.

c) Stouffer's Theory of Intervening Opportunities

In 1940 S.A. Stouffer, a US social psychologist, publicised his **Theory of Intervening Opportunities**, which stated that the number of people travelling any given distance is both directly and inversely proportional to the number of intervening opportunities available to them. By 'intervening opportunities', Stouffer meant job opportunities and the availability of reasonably priced housing. Stouffer's theory can be expressed mathematically as:

$$N_{ij} \propto O_j / O_{ij}$$

where N_{ij} is the number of migrants from town i to town j, $\propto$ is the sign that means 'is proportional to', O_j is the number of opportunities at town j, and O_{ij} is the number of opportunities between i and j.

Stouffer maintained that the nature of an intervening space was a much more important factor than linear distance in influencing migration patterns. The theory of intervening opportunities model has subsequently proved to be a reliable research tool, because comparisons between the predicted and the sets of research data obtained by using its formula have produced consistently high levels of **correlation**.

d) Zipf's Inverse Distance Law

In 1949 G.K. Zipf based his **Inverse Distance Law** on the general principle of **distance decay**, in recognition of the fact that difficulties such as cost and time-wastage invariably face people who undertake long-distance migrations. This principle is incorporated in the **Gravity Model**, in which flows between places are inversely proportionate to the distances that separate them. Zipf's own law can be shown in mathematical form, as:

$$N_{ij} \propto 1/D_{ij}$$

where N_{ij} is the number of migrants moving from town i to town j, $\propto$ is the symbol for 'is proportional to' and D_{ij} is the travelling distance between these two towns.

It is now widely accepted that Zipf's formula is somewhat flawed, because the importance of distance is not so consistently great as to justify a rigid inverse relationship between it and volume of migration. Later models have refined and improved Zipf's formula by adding important migration factors such as relative unemployment rates, per capita incomes and the different proportions of adults of working age between the pairs of towns linked by migration flows. Such refinements to Zipf's Inverse Distance Law also recognise the importance of regular personal contacts between areas of migrant origin and destination; clearly, people who receive detailed, up-to-date information about distant places and have frequent contact with 'significant others' in them are much better placed to overcome any problems caused by separation and distance.

The Gravity Model was later refined by taking into account the 'relative attractiveness' of places of different population size, producing this second mathematical formula:

$$N_{ij} \propto k\, P_i\, P_j\, /\, D_{ij}^2$$

where N_{ij} is the number of migrants moving from town i to town j, $\propto$ is the symbol for 'is proportional to', k is a constant, P_i is the population of town i, P_j is the population of town j and D_{ij} is the travelling distance between these two towns.

3 Pull and Push Factors

One of the simplest and most user-friendly of all migration theories was the series of **Laws of Migration** conceived by Everett Lee in 1966. These were based on what he believed to be the four sets of factors common to all migrations:

* factors linked to the nature of the migrants' destination
* factors associated with the nature of the migrants' area of origin
* intervening 'obstacles', which determine the nature of the journey between origin and destination
* 'personal' factors – those specific to the individual migrant(s) making the journey from origin to destination.

Lees' four laws are shown diagrammatically in the pull and push factor model in Figure 3.1. Lee did acknowledge that **push factors** might be more influential than **pull factors**, if only because individuals have an intimate knowledge of their place of origin and this allows them to make informed, rational decisions. Their judgements of the destination are, however, likely to be somewhat over-optimistic, being based on acquired 'knowledge' instead of direct personal observation. There is still much active debate as to which of the two

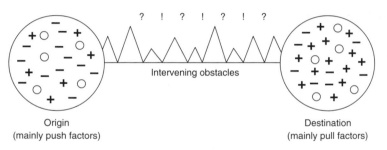

Figure 3.1 Push and pull factors in Lee's Laws of Migration

factors might really be the stronger one, partly because our knowledge of past migrations provides plenty evidence in support of each factor's claim to be the stronger! Every migration is, of course, a unique event due to the infinite number of permutations of individual circumstances within Lees' four factors, and particularly within his fourth (person/group-specific) factor.

Some considerations that determine migration destinations are clearly either pull or push factors (e.g. the presence of a reliable supply of safe drinking water, which can only be regarded as an advantage and so be a pull factor). Other factors, such as those listed below, are less easy to classify because they are dependent on an individual's personal preferences:

- vibrant, urban densely-populated areas *in contrast to* much quieter, sparsely-populated areas
- a warm 'summer' climate *in contrast to* a cooler 'autumnal/winter' climate
- lowland 'prairie' landscapes *in contrast to* rugged, mountainous environments
- nearness to rivers, lakes and seas *in contrast to* 'dry' wilderness areas such as deserts.

Pull and push factors may be categorised in a variety of ways, but the four-group system below is recognised as an effective way of doing this.

a) Physical, environmental and locational considerations

- Naturally occurring events such as floods, earthquakes and tsunamis.
- Climatic characteristics such as droughts and seasonal temperature variations; also, wind-belt and air-humidity patterns, for reasons of personal comfort and peoples' ability to undertake physical work.
- Locations of morphological features such as mountain ranges, major river courses and deserts, all of which represent natural barriers to migration.
- Occurrence of endemic diseases.
- Remoteness and accessibility of places.

b) Economic considerations

- Presence of fertile soils and irrigation water to aid agriculture.
- Changing demand for agricultural workers due to mechanisation and land-use reform.
- Changing demand for workers in the secondary and tertiary sectors of employment due to technological advances (e.g. **automation** in factories).
- The availability of natural resources which can generate power and form the basis of industrial development.
- Overpopulation and underpopulation.
- Employment opportunities such as the nature of the employment available, rates of pay and entitlement to paid holidays.
- Cost of living, especially of basic human needs such as housing, food and clothing.
- Comparative wage levels for similar work carried out in different locations.
- Availability of retirement pensions, sickness benefits, maternity leave, etc.

c) Social considerations

- Crime rates involving people and property; the impartiality of law-enforcement agencies such as the police and the judiciary.
- Location of relatives and friends.
- Availability and quality of housing provision.
- Availability of educational provision ranging from initial schooling to higher education; availability of specialised professional and trade training.
- Availability and range of recreational, social and sporting facilities.

d) Political, cultural and human-conflict considerations

- Political stability.
- 'Freedom of speech', in both verbal and written forms.
- Attitudes towards minority ethnic and religious groups.
- Gender issues, especially attitudes towards women.
- Attitudes towards and controls regulating immigration and political asylum.
- Freedom of movement within a country.
- International/regional conflict and civil war.

e) Personal/family considerations

- Bereavement, marriage and other major personal events.
- Marriage opportunities.
- Promotion and other personal work-related factors.

- Dependency on other individuals who have already taken the decision to migrate.
- Knowledge and perceptions of other locations.

The case study that follows illustrates many examples of the factors listed above and which collectively led to the widespread depopulation of Britain's most northerly and most mountainous rural region over the past 250 years.

CASE STUDY: RURAL DEPOPULATION FROM THE SCOTTISH HIGHLANDS AND ISLANDS REGION

The 'Highland and Islands' region of north-west Scotland is one of Britain's most inhospitable rural areas (Figure 3.2). Migration from this region over the past 250 years has been the result of many different environmental, historical and economic push factors.

Figure 3.2 Typical landscape in the Scottish Highlands and Islands region

a) Physical relief factors

Large parts of this region are very mountainous, with many steep-sided peaks over 1000 m above sea level, including Ben Nevis, Britain's highest mountain (1343 m). During the last Ice Age, glaciers created long, deep troughs, many of them since drowned by inland freshwater lochs and coastal fiords. The lochs have made most agriculturally productive lowland areas inaccessible to the major urban centres further south, because road routes have to follow long detours around their shores; also, few railway lines were built because of the high cost of constructing the many bridges and tunnels needed. Some of the larger inhabited islands are separated from the mainland coast by dangerously strong tidal currents. Glaciation processes also eroded and transported away much valuable topsoil, leaving shallow layers often made even more difficult to plough by exposed outcrops of solid rock. Much of the remaining soil is water-logged or heavily leached acidic, peaty soils.

b) Climatic factors

The region has a total annual precipitation above 2000 mm in most areas due to both **orographic** (relief) and **frontal** (cyclonic) factors. It is exposed to the prevailing south-westerly winds, which have a high moisture content following evaporation of the relatively warm water of the North Atlantic Drift ocean current. Low air pressure **depressions** migrate eastwards across the British Isles, after having crossed the Atlantic Ocean, creating bands of heavy rainfall along their **warm**, **cold** and **occluded fronts**. The moisture-laden air rises above the mountains upland and cools to form rain, sleet and snow. Winter snowdrifts may block vital road and railway routes, and deny remote settlements access to essential services such as ambulances.

c) Historical and economic factors

Migration from north-west Scotland has only partly been due to the challenges presented by its hostile natural environment. In 1745 'Bonnie Prince Charlie' led an uprising of some of the Highland clans, the purpose being to invade England and then depose its king. His invasion army of volunteer Highlanders was no match against King George II's professional soldiers and it suffered a crushing defeat at the Battle of Culloden. The victorious English treated any survivors extremely harshly and made it unlawful for Scots to carry weapons or even wear their clan tartans. As a result of these measures, the chiefs' hereditary powers became so weakened that they could no longer prevent their clansmen from leaving the villages. Thousands took advantage of this unexpected freedom and migrated to the Scottish lowlands to work in Clydeside's rapidly expanding coalmining, iron and steel, shipbuilding and textile industries. The chiefs' incomes from their remaining clansmen were reduced so much that they were forced to seek alternative ways of earning money. Sheep rearing appeared to be the ideal solution, because sales of lamb and wool had already been shown to be far more profitable than the clans' traditional occupations of farming, fishing and weaving. To accelerate the migration process, some chiefs employed ruthless tactics such as the burning of Highlanders' croft-houses. This forced migration, known as the Highland Clearances, lasted from 1785 to 1850 and many of the reluctant migrants sailed to British colonies in Australia, New Zealand and (especially) North America.

In fact, the Highlands and Islands were already seriously over-populated well before these voluntary and enforced migrations took place. The high rate of depopulation continued well into the mid-twentieth century and created serious economic and social problems for the region that continue to the present time.

Between 1961 and 1981, 27 Scottish islands became totally depopulated. An island which lost many of its people due to emigration is Lismore – one of the lowest lying and most fertile of all of the Scottish islands. Its largest population, according to the 1831 census, was 1790; its current population is only 156, fewer than 9% of this peak total. Evidence of prolonged depopulation can be seen throughout Lismore, which is littered with ruined, abandoned crofts. One of the island's few remaining population clusters is Port Ramsay, once a thriving farming and fishing community, but now most of its tiny cottages are holiday homes and none of its permanent residents is under 65 years of age. One of the island's two schools closed in the 1960s and all its older pupils have to live in Oban on the mainland during term-time. Not surprisingly, these young people become very self-reliant and quickly gain the confidence to relocate further away in search of well-paid jobs with good promotion prospects. Students wishing to undertake degree courses or advanced professional training have no alternative but to migrate to cities such as Edinburgh and Glasgow. However, a recent boost to Lismore's economy was the opening of nearby Glensanda 'super-quarry', which extracts seven million tonnes of high-quality granite annually and currently provides six of the island's residents with full-time, pensionable employment.

The graph in Figure 3.3 shows that recent population trends within the Highlands and Islands have been somewhat inconsistent. This is because the success of individual settlements depends

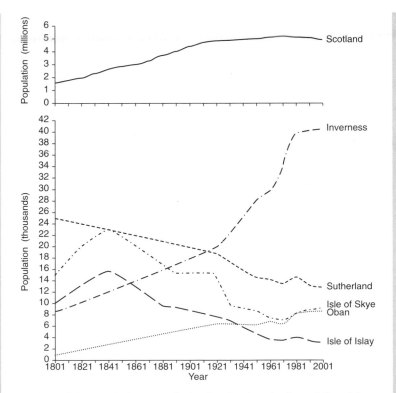

Figure 3.3 Population trends for Scotland and selected Scottish
regions: 1801–2001

very much on their own location, accessibility and the ability of
government-funded organisations such as the Highlands and
Islands Development Board (created in 1965 and replaced in
1990 by Highlands and Islands Enterprise) to create employment
opportunities. In recent years, the European Union has allocated
significant development funding to its **peripheral regions** experiencing **socio-economic decline**. Skye's population increase
has largely been the result of the modern road bridge that links
it to the mainland and has enabled the island to attract over
600,000 tourists each year. Thousands of construction and main-
tenance workers were based in the Orkney and Shetland Islands
following the discovery of North Sea oil in 1970. Long-established
towns such as Oban have prospered due to tourism and their key
roles as **nodes** in the region's road and heavily subsidised rail
and island ferry routes. Inverness is the 'capital' of the Highlands
and Islands Region and many of its people are employed in local
government administration. Whisky distilling, electricity generation and aluminium smelting continue to be important sources

of employment, even though they are labour-intensive only during the initial construction phase. The Forestry Commission has planted large numbers of trees on mountainsides and in the glens (valleys) since its formation in 1919. More recently, fish farming in sheltered sea lochs has expanded rapidly and Scotland is now Europe's foremost producer of farmed salmon.

4 The Impact of Migration

The impact of migration on communities depends on three factors:

- the relative size of the communities involved
- the rate of migration
- the proportion of the total populations involved in both the donor and receiver locations.

Total migration results in entire communities ceasing to exist, at least for a period of time, although re-inhabitation may take place if circumstances change sufficiently for this to happen. Most of the land on the Maldives' 200 islands is only one metre above the level of the Indian Ocean and three of those islands have already been totally and permanently evacuated as a result of sea level rises due to **global warming**. Some oceanographers predict that all the remaining islands in the Maldives **archipelago** (group of islands) will be also become uninhabitable within the next 30 years.

Two examples of short-term total depopulation involve the remote islands of St Kilda and Tristan da Cunha. In the case of St Kilda – 65 km from the nearest Outer Hebridean island of Benbecula – the quality of life of its inhabitants had deteriorated so severely that its residual population (36 old and infirm people) was persuaded to relocate to the Scottish mainland, in 1930, after receiving assurances that they would all be housed. Military **reinhabitation** of St Kilda followed the construction of a missile-tracking station on the island in 1957. Tristan da Cunha – one of the world's most remote communities – lies in the South Atlantic Ocean mid-way between Africa and South America, 2800 km from Capetown. It remained uninhabited during 1961–3, when all 268 of its people were evacuated following predictions of what later proved to be very dangerous volcanic activity.

a) The effects of out-migration

Partial migration often reduces the quality of life of the **residual population** because the viability of both private businesses and vital public services is threatened. Any closures of schools, shops, post offices, libraries, banks and garages – as well as medical, transport, emergency and entertainment facilities – often accelerate the rate of emigration. Significantly, £65.9 million of donations were made to

British village halls and community centres during the National Lottery's first 10 years. In very small coastal communities, reductions in the number of able-bodied adults may make it impossible to crew fishing boats. Over time, the financial burden of subsidising remaining essential services may become too much for the residual community. In such cases, total depopulation may occur unless additional funding can be made available by external agencies such as government departments and international organisations. An example of such intervention is the EU's extra funding for its peripheral regions such as the Shetland Islands, whose primary schools are now some of the best equipped in the whole of Scotland.

House values may fall significantly due to the reduced demand for accommodation. Occasionally, the migration or retirement of just one individual can traumatise an entire community. On the Inner Hebridean island of Gigha, the retirement of Seumas McSporran, MBE, from his 16 part-time jobs including postmaster, undertaker, registrar (of births, marriages and deaths), coastguard, fireman and policeman posed considerable difficulties for the island's community of some 120 people! Some Hebridean islands have now taken the initiative to reverse their depopulation trends. Gigha has recently built 20 new houses, most of which are only available for rental and only by incomers with young families. When a nearby island called Muck recently advertised in a national newspaper for 'new blood', over 200 families applied to migrate there! Six of them were shortlisted and invited to spend a week on the island as a trial period. The 22 adult inhabitants of Muck finally chose a family from Leeds, which had two young daughters and whose mother had crucially just applied for the vacant post of island primary school teacher.

Out-migration can jeopardise future business investment in a community, partly because it creates uncertainty about the availability of suitable workers. It is usually the younger, more active and more ambitious adults with young families who take the bold decision to leave their rural areas, leaving a residual population consisting mainly of older, retired people. The migration of farmers often leads to cultivated land reverting to unproductive wilderness.

b) The effects of in-migration

In-migration can have many beneficial effects on receiver communities, whose economic potential and cultural vitality can be greatly enhanced by it.

Professional and technically skilled immigrants have the ability to improve the quality of local public services. First-generation migrants are generally much more willing to accept jobs in the less-skilled, 'dirty, dangerous and difficult' jobs that the indigenous people are increasingly reluctant to accept. A plentiful supply of workers in the more labour-intensive industries keeps wage levels down, reduces manufacturing costs and makes manufactured goods more competitive on international markets.

Immigrants are particularly welcome if they can fill short-term needs, and it is not unusual for countries to advertise overseas for workers in skills-shortage areas. For example, the usual annual quota of immigrants permitted to enter Britain specifically for work on fruit and flower farms was about 18,700, but this figure was suddenly increased by 6000 in 2002 – simply because one-third of the previous year's crop of Cornish daffodils (valued at £3 million) had rotted in the fields due to a severe shortage of flower-pickers. Migrants are especially welcome when a country's economy is expanding, but unfortunately often begin to experience hostility when job vacancies decrease, unemployment increases and indigenous workers accuse incomers of 'taking their jobs'. In the short term, migrants tend to be a burden on the host country's resources – particularly welfare support budgets – due mainly to their additional child-support and medical requirements. It is a considerable challenge for them to obtain worthwhile jobs – often because of language difficulties – and this leads to them becoming temporarily dependent on unemployment benefits. A high proportion of first-generation migrants do find life very difficult, especially if they have entered the country illegally. A tragic example of this occurred early in 2004, when 20 Chinese illegal-immigrant cockle-gatherers were drowned on sandbanks in Morecambe Bay by the incoming tide while working in a very hazardous environment for rates of pay well below the **National Minimum Wage**.

Second and subsequent generation migrants tend to be more familiar with the host country language, be more successful at school and contribute towards to the national economy due to their increased tax payments and reduced reliance on family support. In 1999–2000, the UK total migrant population contributed £31.2 billion in taxes, but consumed £2.4 billion less than this in state benefits.

Not all immigrants are low skilled; many are professionals whose knowledge and expertise is of great benefit to the recipient country. Particular skills shortages can be alleviated, as when Britain issued more than 6000 non-EU work permits to teachers to compensate for serious shortages in the classroom. In 2001 alone, over 600 Jamaican experienced teachers were lured into UK schools by financial incentives. In 2002, 50,000 foreign nurses – mainly from countries in sub-Saharan Africa – responded to advertised incentives to work for the National Health Service. These appeals proved so successful that, two years later, some African countries pleaded with Britain to reduce its overseas recruitment of nurses because it was having such an adverse effect on AIDS treatment in Africa.

Cultural exchange is potentially enriching and can lead to much greater diversity in catering, the arts and many areas of public life. One sector of the British community is particularly enhanced by recent in-migration: the Church of England experienced a decline from 1.4 to 1.14 million in attendance at services between 1990 and

2004 (a fall of 18.6%), but the membership of Afro-Caribbean churches in England almost doubled from 60,000 to 115,000 during the same period! In-migration can strengthen a host country's relationships with its trading partners, partly by increasing its number of translators but also by making its representatives more aware of local business etiquette and social customs.

Migrants represent a potential future increase in national earning power, particularly for those countries whose growth rate is consistently below the minimum rate of 2.1 children per woman needed to maintain a stable population. The EU's current growth rate is only 1.65 and its working population is predicted to decline after 2010, whereas that of the USA will accelerate; substantial migration would help to make the EU more competitive by compensating for its negative population growth rate.

5 Intercontinental Migration

Migration between continents has been an increasingly important feature of population movement since the end of the Ice Age, some 14,000 years ago. One of the most outstanding examples remains the forced migration of Africans to the Americas during the Slave Trade of the sixteenth to nineteenth centuries. This infamous trade created great wealth for the maritime nations of Western Europe and the transportation of slaves was only one of the three highly profitable legs of the trans-Atlantic 'trade triangle'. This pattern of trading began in a very modest way, when an enterprising merchant bought a small number of prisoners from an African tribal chief in 1518, then sold them at a substantial profit after transporting them across the Atlantic. The Slave Trade resulted in the widespread depopulation of many European colonies in West Africa and only ceased within the British Empire when it was declared illegal in 1833. Although well over 10 million slaves were transported in this way, not all of them reached their destinations. Of the three million transported between 1666 and 1776, no fewer than a quarter of a million died during the 6–8-week ocean voyage. The populations of some of the Caribbean islands were transformed by this rapid influx of newcomers, Jamaica alone receiving 610,000 slaves during the period 1680–1786. Figure 3.4 maps the Slave Trade and other early examples of intercontinental migration. Trading in slaves still remains active and profitable in certain parts of the world, particularly between some of the countries bordering on the Indian Ocean and its adjacent sea areas. In West Africa, slavery continues to the present day in a form that is even more repulsive than before because it involves children. Anti-slavery International estimates that about 200,000 children – some only four years old – are bought in for as little as £10 each in the poverty-stricken villages of Benin, Mali and Togo, then sold to cocoa plantation owners in the Ivory Coast or shipped across the Gulf of Guinea to work as domestic staff in the prosperous homes of Gabon, whose economy has been boosted by its thriving tourist industry.

CASE STUDY: MIGRATION INTO BRITAIN

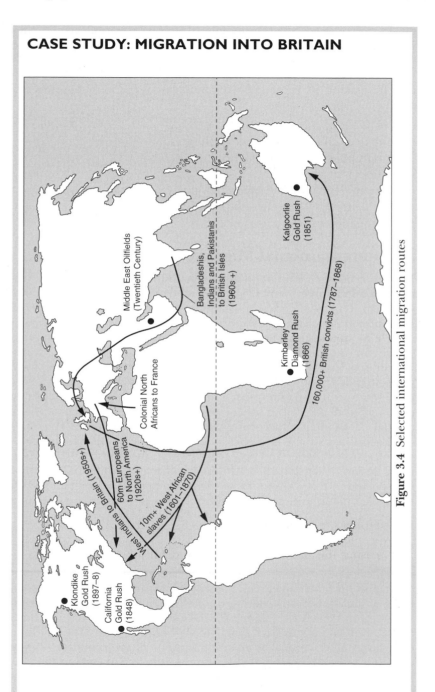

Figure 3.4 Selected international migration routes

a) Population composition

Figure 3.4 also displays some more recent intercontinental migration flows, most of which have been undertaken voluntarily and for personal economic reasons. Many of these migration flows have had a considerable impact on the composition of Britain's population, especially as immigrant ethnic groups continue to exhibit much higher birth rates than those of the indigenous population. Figure 3.5 displays the UK's current population composition as revealed by the 2001 national census and contrasts this with the equivalent situation 50 years earlier (although strict comparisons are difficult because pre-1991 censuses requested 'country of birth', whereas the 1991 census introduced a 'membership of ethnic group' question). The populations of donor countries have also been affected quite substantially. Between 1845 and 1932, 50 million Europeans emigrated to North America and Australasia, 20% of Europe's natural population increase, and the modern equivalent of about 80% of the current total population of France.

b) Migrant origins

People have been migrating into Britain for thousands of years, but the rate of immigration increased sharply during the last two centuries due to rapidly improving international transport networks that made it possible for large numbers of people to travel much longer distances. Many Irish people arrived during the nineteenth century to obtain work building canals and laying railway tracks. Few 'non-white' immigrants entered Britain during the 1800s, although some Chinese did establish laundries and, much later, food 'take-aways' that thrived due their competitive prices and high standards of service.

Immigration into Britain peaked after its economy had recovered following the devastation of the Second World War. Post-war governments had decided to encourage immigration as a way of increasing the national workforce and fill job vacancies in the less-skilled occupations. It placed newspaper advertisements in India, Pakistan, the Caribbean islands and many other British overseas territories; financial assistance was offered to those willing to migrate. In 1956, London Transport placed similar adverts and offered to loan the cost of the transatlantic fare, which would be repaid later on by deductions from wages. These advertising campaigns proved highly successful and produced a net migrant gain of almost 400,000 during 1960–2 alone. In fact these campaigns were so successful that the government felt it necessary to pass the **Commonwealth Immigration Act** in 1962 to

Census year	Total UK population (in millions)	Total of ethnic minority population (%)	Individual ethnic minority groups (%)					
			Black (African)	Black (Caribbean)	Bangladeshi	Indian	Pakistani	All others
1951	48.8	0.45	0.03	0.03	c0.01	0.25	0.02	0.11
2001	58.8	7.99	0.90	1.04	0.52	1.94	1.38	3.21

Figure 3.5 Changing ethnic composition of Britain's population: 1951–2001

restrict further immigration from these countries. By the late 1960s, the number of immigrants had fallen to less than 30,000 per year – many of these being relatives of immigrants already living here. Britain was finally beginning to be recognised as a truly multiracial society and the **Race Relations Acts** were passed in 1965 and 1968 to offer increased protection to its expanding ethnic minority groups. These acts of parliament made it an offence to discriminate against people purely because of their colour, religion or cultural differences.

c) Migrant concentrations

Newly arrived immigrants have always tended to settle in the largest towns and cities. In the late eighteenth and early nineteenth centuries so many Italian immigrants lived in the Hatton Garden and Clerkenwell districts of London that they became popularly known as 'Little Italy'. During the nineteenth century, Irish immigrants were particularly attracted to the west coast ports such as Liverpool and Glasgow, because they were within easy reach and offered plentiful manual employment in their docks, processing industries and transport **infrastructures**. In the 1930s, European Jews escaping Nazi persecution adopted the Golder's Green district of north London as their main 'base' in southern England; 70 years later, the Jewish community remains one of Golder's Green's chief population components. Post-war immigrants from the British Commonwealth have similarly been drawn to the major cities. Southall, also in London, was one of the first districts to be settled by this later wave of immigrants, which included many people from the Indian sub-continent. When the Caribbean Islands passengers on the liner *Empire Windrush* landed at London in the summer of 1948, many of its 492 passengers settled in Brixton – simply because the nearest job centre to the docks happened to be there. The growing popularity of the Notting Hill Carnival, held annually in north London for many years, is evidence of that area's rich ethnic community originating mainly from Barbados and Trinidad. Figure 3.6 locates all these as well as other centres of immigrant population within the Greater London conurbation.

London continues to be home to the UK's largest ethnic communities and research undertaken in 2000 by Dr Philip Baker of Westminster University revealed that more than 300 languages were then being spoken by London schoolchildren; it also indicated that only two-thirds of the capital's 850,000 children regularly spoke English at home. The northern city of Bradford now has one of our highest concentrations of Muslim

Pakistanis as well as the UK's lowest proportion of indigenous, white people (80%).

Cities have always held many attractions for new arrivals. They are major centres of employment and so are most likely to provide plentiful job opportunities that are accessible to first-generation immigrants. Very important too is the availability of reasonably priced accommodation. During the 1960s, British cities still had large numbers of nineteenth-century terraced houses within their inner residential areas, conveniently situated close to the central business district, dockland and industrial zones.

There are many 'social' reasons why immigrants tend to live close to other members of the same ethnic group. Friendship ties give them an increased feeling of personal security during the first few crucial years following arrival. Relatives are available to offer advice and possibly financial assistance and temporary accommodation. Younger adult immigrants are more likely to find suitable marriage partners within their ethnic clusters and

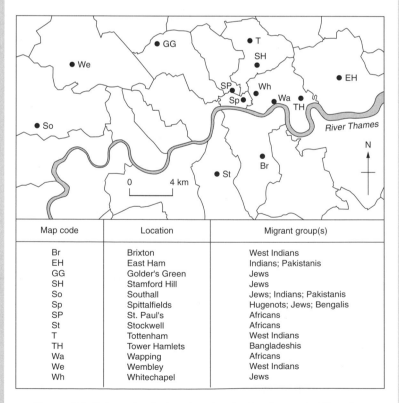

Map code	Location	Migrant group(s)
Br	Brixton	West Indians
EH	East Ham	Indians; Pakistanis
GG	Golder's Green	Jews
SH	Stamford Hill	Jews
So	Southall	Jews; Indians; Pakistanis
Sp	Spittalfields	Hugenots; Jews; Bengalis
SP	St. Paul's	Africans
St	Stockwell	Africans
T	Tottenham	West Indians
TH	Tower Hamlets	Bangladeshis
Wa	Wapping	Africans
We	Wembley	West Indians
Wh	Whitechapel	Jews

Figure 3.6 Selected ethnic minority clusters in central London

social facilities that are compatible with their cultural expectations. Religious facilities tend to be particularly important to 'homesick' newcomers who are in need of reassurance from respected elders from their traditional societies, and larger ethnic communities are better able to provide special buildings such as mosques for their own use.

Unfortunately, inter-racial harassment is also a location factor for the more vulnerable members of minority groups. Living within a cluster of people from one's own background does offer some measure of protection and support when racial issues escalate. Deteriorations in inter-cultural relations can occur quite quickly, but are most predictable during times of economic recession when job opportunities are reduced and aggrieved, and unemployed members of the indigenous population vent their frustration on those members of ethnic groups who do have jobs. The clustering of ethnic groups enables those who perceive themselves to be disadvantaged to take mutual protective action during times of peak stress – such as the riots that occurred during the summer of 1981 in the Toxteth area of Liverpool, the St Paul's area of Bristol and other inner-city areas, with smaller-scale repetitions in 1985. Racially motivated disturbances also occurred on a reduced, but still significant, scale in July 2001 in the northern England industrial centres of Bradford, Burnley, Oldham and Leeds.

In late May 2002 the Belgian city of Antwerp was the scene of vicious assaults by Arab youths on Jewish families dressed in their distinctive black high boots, coats and fur hats. Personal attacks such as this and damage inflicted on Jewish synagogues became so common that Belgian newspapers actually stopped reporting them – they were simply no longer regarded as 'newsworthy' events. France is also one of the many European countries to have experienced racially motivated disturbances, particularly with respect to its migrant population from North Africa with whom it used to have very close colonial links.

6 Transmigration

Transmigration is the circulation of people *within* a country and, as in the case of intercontinental migration, it has taken both 'forced' and 'voluntary' forms. The following two case studies provide recent examples of both.

CASE STUDY: FORCED TRANSMIGRATION IN INDONESIA

Events in Indonesia in the second half of the twentieth century have provided one of the classic examples of what is known as **forced transmigration**.

a) The historical background to the Indonesian forced transmigration

Indonesia has a very uneven distribution of population. Its most densely populated areas are on the more centrally located Indonesian islands of Java, Bali and Lombok, where rich volcanic soils have sustained a centuries-old system of highly intensive subsistence agriculture. The nation's capital and **primate city** (most highly populated settlement), Jakarta, is located at the northwestern coastal tip of Java.

Transmigration is not a new component of Indonesia's population and economic policies. It was introduced by the Dutch as long ago as 1905, during their colonial rule, and was later adopted by the Indonesians after their country became independent immediately after the Second World War. Resettlement began in 1905 under Dutch colonial rule, but the rate of population movement accelerated following independence. The policy was further strengthened in 1969, four years after the start of President Suharto's 33-year dictatorship. Suharto's own brand of transmigration was, however, very much of the 'forced' type and involved widespread brutality. It resulted in the annexation of West Papua and an invasion of East Timor.

b) The nature of the forced transmigration

The basic concept was to use transmigration from the core islands of Java, Bali and Lombok as a mean of exploiting the potential of the peripheral (outer) islands; these included Sumatra, Sulawasi, and the Indonesian territories of Kalimantan (part of Borneo) and Irian Jaya (part of New Guinea, and since renamed as West Papua by the Indonesian government for political reasons). This exploitation was to be achieved by stimulating economic and infrastructural development as well as providing land and new opportunities for some of Indonesia's many destitute and landless families. It was also used as a political tool to control the indigenous populations of these outer areas and crush any hopes they might have of becoming self-governing, independent countries. Because of this, there was no prior consultation with the local populations and it was, therefore, almost inevitable that transmigrants and local people would clash with each other.

Australia and the USA had some sympathy with the Indonesian government's desire to eliminate groups of independence seekers, because both countries had recently suffered a humiliating military defeat in Vietnam and had no wish for the political *status quo* in south-east Asia to be undermined by similar ambitions within Indonesia. In 1975 East Timor was seized by the Indonesian army to provide additional land for transmigrants from Java. Only three years later, and after substantial deposits of oil and gas had been discovered under the Timor Sea, the Australian government became the first in the world to formally recognise Indonesia's annexation of East Timor. It is quite probable that its motive for doing this was to establish an early claim to a sizeable share in the untapped wealth below the Timor Sea!

Between 1949 and 1974 the Indonesian government resettled 674,000 people through transmigration. Another 3.5 million people were resettled to transmigration sites on the outer islands by 1990. A total of six million people moved during the entire period of the forced transmigration (see Figure 3.7 for details of the changing migrant flow at five-year intervals).

At least 200,00 East Timorese perished during their ruthless suppression by Indonesian forces. As a result of conflicts such as these, thousands of transmigrant families had no option but to become refugees and flee to other areas. Often, the number of

Migration period	Number of migrants
Up to 1969	500,000
1969–74	174,000
1974–9	544,000
1979–84	2,469,560
1984–9	1,061,680
1989–94	no reliable data
1994–9	1,500,000
1999 and after	22,000

Figure 3.7 Transmigration flows in Indonesia from 1969

transmigrants was far exceeded by the number of refugees, who then formed over 25% of the total population of some of the outer islands.

The transmigration programme received considerable **multilateral aid** from the World Bank as well as **bilateral aid** from individual countries, the latter making the recipient vulnerable to agreements, which are usually heavily biased towards the donor's best interests. During the 1990s, the focus of overseas financial assistance changed to supporting the rehabilitation of existing resettlement projects instead of stimulating fresh waves of transmigration.

c) The consequences of the forced transmigration

- Because many transmigrants came from cities and lacked any basic knowledge of cultivation techniques it is hardly surprising that a high proportion of the new agricultural initiatives failed almost immediately. Some of the new settlements were sited in peat forests and tidal swamps in Kalimantan, where traditional agriculture is challenging even for the most experienced subsistence farmers. In East Timor, which became fully independent of Indonesia in 2002 after 25 years of armed conflict and chose to be known as the Democratic Republic of Timor-Leste, more than 40% of its people now live below the poverty line, over half are illiterate and at least 50% of infants are underweight. Life expectancy in East Timor is only 56 years, malnutrition there is endemic, and diseases such as malaria, denge fever and tuberculosis are now rife due to the rural population's rapidly deteriorating quality of life.
- Those islands that have received transmigrants contain the majority of south-east Asia's remaining primary forest, which is second in size only to that in the Amazon Basin, and is one of the world's most ecologically diverse regions. It is also the only global habitat of some 100 species of mammal. Deforestation carried out to create extra farmland for the migrants destroyed much of this invaluable global resource. Over 50 million hectares of forest have been cleared in the past 40 years for logging and agricultural development, and the country's current *annual* loss of primary forest is estimated at 1.2 million hectares. Soil erosion is widespread and increased leaching further damages the topsoils and seriously reduces their fertility.
- The programme was an economic disaster for Indonesia and greatly increased its national debt. Between March 1998 and March 2000, this increased from 23% to 91% of the country's total GDP.

- The extent of the failure of the forced transmigration programme is evident by the large number of migrants who are
now returning to the overcrowded cities within the core
islands of Indonesia.
- These disastrous events in Indonesia have prompted a number
of **non-governmental organisations (NGOs)** such as Oxfam
to undertake research into transmigration, a form of
migration that had previously attracted far less interest than
other types of population circulation.

CASE STUDY: VOLUNTARY TRANSMIGRATION IN POST-WAR GERMANY

Germany had the largest population of any Western European
country before its separation into East and West Germany at the
end of the Second World War.

The so-called Cold War situation that existed between the
Communist East and the non-Communist West prevented transmigration between the two countries until the re-unification of
Germany in 1990, to the great delight of both former countries.
During the Cold War period, West Germany's economy had
flourished, so much so that additional labour was needed to sup-
plement the indigenous workforce. This boost to the German
workforce was achieved by means of voluntary migrations from
Turkey and the poorer countries in Southern Europe. One of
the recommended research projects at the end of this chapter is
based on the nature of that economic migration and the so-
called *gastarbeiter* (translated as 'guestworkers') who took part
in it.

It was inevitable that post-reunification transmigration would
take place between the two German countries. The infamous
Berlin Wall (which was built by the Communists to seal off East
Berlin from the more prosperous western side of the city) had
separated many families and its destruction just before reunification took place meant that people could visit their relatives for
the first time in decades. East Germany's economy had stagnated
in comparison with that of the West, where new post-war factories
equipped with modern machinery permitted the mass production of goods at very competitive prices; the result was increased
wages and a very high standard of living.

Many East Germans transmigrated for purely economic
reasons, but younger adults were also attracted by the vibrant

social life that was enjoyed by their contemporaries in the West although, ironically, today former East Berlin has a much more vibrant 'scene' than former West Berlin!

The following newspaper article shows the effects of 15 years of transmigration on Wittenberge, a typical, medium-sized town in the centre of the former East Germany.

East Germany's vision of unity fades (**Daily Telegraph,**
11 September 2004)
Visitors who arrive by train in Wittenberge emerge into a modern £52 million railway station with undulating roofs and glass lifts. The centre of this town is equally impressive. Newly cobbled streets and redbrick cycle paths are just some of the evidence of a £850 billion investment – equivalent to more than five times the GDP of Switzerland – that has been pumped into the East since the fall of the Berlin Wall.

But behind the gleaming elegance of this town, which boasts a far more modern infrastructure than that of any similar-sized community in the West, there lies a much darker side. Since reunification, in an exodus the like of which has not been seen since the Thirty Years War in the seventeenth century, Wittenberge's population has shrunk at an alarming rate from 30,000 to just over 20,000 as the younger citizens have left in desperation to find work elsewhere. More than a decade after Chancellor Helmut Kohl promised the 15 million easterners 'blossoming landscapes', unemployment is now more than 20% – more than twice the German national average. Some experts say it could actually be twice as high as that.

One out of every three Wittenbergers is now aged over 60, and the local maternity hospital was demolished last month to make way for yet another old peoples' home. The town is, quite literally, dying. Wittenburge has seen all its main industries close down over the last decade, including its sewing-machine factory, the spun-rayon factory and the oil-rape-seed processing mill. The people here feel like they are a failed social experiment and 75% of East Germans now consider themselves to be second-class citizens.

Dissatisfaction is also growing among West Germans. In addition to the compulsory basic national taxation, *all* Germans now pay an extra 5.5% of their wages in the form of a solidarity tax towards subsidies to fund the East. Amazingly, a poll undertaken in 2004 revealed that 25% of westerners and 12% of easterners actually wanted the dividing wall between East and West Germany to be re-built!

7 Retirement Migration

Most working people have a very clear idea of where they would like to live and what they would prefer to do after retiring. Many people are content to relocate to Britain's most popular retirement resorts along the English Channel coast, in North Wales and in the Lake District. The more adventurous look abroad, usually after holidaying successfully in their preferred retirement location for several years. Relocating to a foreign country can be a daunting experience and many reduce the risks involved by migrating for a trial period in rented accommodation. Others migrate seasonally, going to a warmer region during the winter months and offer their houses abroad as holiday accommodation while they are back in Britain.

A recent survey showed that the number of UK pensioners living abroad has now passed the one million mark (almost 10% of the pensioner population), compared with only 252,000 in 1981. Italy, France and Spain were identified as the most popular European destinations, while Australia, Canada, New Zealand and Florida are the most preferred distant locations. About 88,000 UK pensioners live in Ireland – one-fifth of that country's own pensioner population – partly because there are no language difficulties or migration restrictions between the two countries. Some of the most important reasons for this increasing trend towards international retirement migration are listed below:

- increasing affluence
- the availability of both state and privately funded pensions
- cheaper air flights, allowing more frequent return trips to visit families and friends in the UK
- the increasing number of UK estate agents specialising in overseas property sales
- the popularity of overseas developments that are restricted to families having members over 45 years of age, so guaranteeing 'quiet havens' for older people
- the rapid increase of suitable new-build properties in the most popular overseas retirement locations
- the trend towards earlier retirement, which is now beginning to ease as governments accept that requiring people to work longer is a key strategy in funding the increasing burden of post-retirement pensions
- the popularity of television programmes and newspaper/magazine articles that highlight the attractions of retiring abroad
- increased longevity, making it worthwhile for older people to become more adventurous when planning retirement arrangements
- discontent with the British weather – in spite of growing evidence that global warming may be 'improving' our summers!

- concerns about issues of personal safety and the security of property at a time when domestic crime rates remain high
- frustration with traffic and pedestrian congestion; also urban pollution due to exhaust fumes, increasing noise and litter proliferation
- adverse publicity about hospital waiting lists and the quality of medical and dental provision at home.

Key factors for selecting a location abroad include:

- the language
- local rates of income and death duty taxation
- house prices and house renovation costs
- the cost of living – especially food, drink and motoring
- nearness to the sea – the British are an island race and are instinctively attracted to maritime locations
- possibly the most important factor of all – the predictability of warm, dry weather.

CASE STUDY: UK RETIREMENT MIGRATION TO OTHER EU COUNTRIES

Recent polls indicate that 54% of the adult British population would like to settle abroad and a quarter of a million Britons actually do emigrate every year, the largest group choosing Spain as their preferred location.

There are many reasons that help to explain both of the above statements. The British climate is notoriously unpredictable and its winters can make life difficult for elderly people. It is, therefore, hardly surprising that over half of all adults in British would opt to live in a much warmer, drier climate if they had the money to do so. Also, older people now have increasing worries about their own personal safety and that of their property. They are aware that Britain has the highest prison population within Europe (as a percentage of its total population). British society is currently undergoing many fundamental changes and these too are of concern to older people, who tend to be very traditional in their attitudes towards community life. The above are some of the more obvious push factors – problems that stimulate out-migration.

There are probably more pull factors – the perceived advantages – that attract people to Europe. EU citizenship allows unhindered travel and residency within an international community whose total populated exceeds 450 million following its enlargement on 1 May 2004. EU citizens can live and work anywhere within its member countries, and when the pound sterling is much 'stronger' than the euro and other mainland currencies; this gives the British additional buying power within these other countries. Also, the relentless rise in UK property prices has enabled Britons to generate substantial cash windfalls that can be re-invested in properties overseas. The ability to keep in regular touch by e-mail and mobile phone has made family separations easier to bear. The British are often thought to be very 'insular', but history shows that they have always been one of the most willing of nationalities to emigrate. Many European countries have a more relaxed life-style and their taxation levels are generally much lower. Those who wish to work abroad before finally retiring can expect to work fewer hours per week on mainland Europe.

The following summaries indicate the range of the environmental, financial and cultural factors that, collectively, make Western and Southern Europe so attractive to retired British people.

* Spain has a typical **Mediterranean Region climate** with distinct seasons composed of very warm, dry summers and mild, wet winters. In Spain, homes even in the most popular areas cost only about one-third of those in Britain. In 2001 alone, 150,000 British migrants bought homes in Spain, which has recently overtaken France as the most popular retirement destination. About 750,000 homes in Spain are now owned by British people, compared with 500,000 in France. The decreasing numbers of tourists visiting Spain has helped to keep house prices there at modest levels and many 'expats' are able to isolate themselves by creating replica, traditional English communities.

* France is Britain's nearest European neighbour and it is for this reason that many schools organise educational visits there to support their foreign language teaching. Adults visiting as tourists are often surprised by how much 'school French' they have remembered and are aware of how using the local language can help them to integrate more quickly into their new local communities. One of the most popular tourist and retirement regions in France is the Dordogne, which has many natural and human features attractive to British people. These include rivers such as the Dordogne, from which the region takes its name, and which flow in steep-sided valley gorges. There are many villages built of local stone, fortified medieval towns and castles. In the Dordogne town of Bergerac, 15 of the 75 pupils in one of the local primary schools are British, some of them having parents who are now able to commute to their workplaces in Britain using the town's regular, budget air flights. Cross-Channel commuting is also proving increasingly

popular with Greater London workers who have bought properties in the Pas de Calais region and use the Channel Tunnel's 'Eurostar' trains in preference to the crowded south-east England commuter rail network. Many later retire in France, having established themselves within the local communities and become genuine 'Francophiles'.

- The Maltese islands enjoy similar weather to Spain's Costa del Sol, but have a somewhat less vibrant tourist social life. Both Maltese and English are the official languages in this former British colony and the modern independent state is a member of the British Commonwealth. Malta abolished its inheritance tax in 1992, replacing it with a modest 5% wealth transfer tax on death. This has encouraged many wealthy people to buy properties in Malta. Housing represents very good value for money and it is possible to buy a 12-room town house in Sliema for the equivalent cost of a 1930s three-bedroom semidetached house in a typical London suburb.

Summary

- Migration can take place at a range of scales and for a variety of reasons: local–intercontinental; forced/voluntary; seasonal–permanent; rural–urban; for economic benefit/to avoid natural disasters.
- A wide range of migration theory has been undertaken by: Lees, Ravenstein, Zelinsky, Stouffer and Zipf.
- The effects of emigration may include: loss of young, able, dynamic and healthy people; residual population becomes imbalanced due to dominance of elderly; important community services (e.g. schools) are no longer viable and close down.
- The effects of immigration may include: gain of young, energetic people; many immigrants are skilled workers/professionals who can fill shortages (e.g. in hospitals); recent immigrants tend to be more willing to undertake unattractive work/shift-work; increased racial tension.
- Many past migrations have been forced, e.g. the Slave Trade; others have been triggered by the discovery of precious raw materials (e.g. the Californian gold rush).
- Past emigration from Britain included farmers and administrators to the colonies of the British Empire; the descendents of about 50 million European migrant families now inhabit North America.
- Post-1950s British migration has been dominated by ethnic groups from former colonies such as India, Pakistan and the West Indies.
- Immigrant groups tend to cluster together for a variety of reasons: language difficulties; availability of ethnic group support and services (e.g. mosques and specialist food shops); mutual protection against racial harassment.

- Such groups often cluster in the inner residential zones of large towns and cities, where accommodation is cheaper and near to job opportunities.
- Transmigrations (migrations within countries) are very common, e.g. forced transmigration in Indonesia and voluntary transmigration in Germany.
- Retirement migration is on the increase, due to greater affluence and longevity; coastal areas within the UK and many EU countries are particularly attractive locations, although rural areas with character and significantly lower house prices are also popular.

Student Activities

Research projects

1. The case study on pages 95–96 provided a considerable amount of information about post-reunification transmigration from the area that used to be East Germany. Before that migration flow had taken place West Germany had relied on international migration to fill gaps in its workforce, in the form of 'guest workers', economic migrants who were attracted from the less economically developed Mediterranean countries such as Turkey. Using the Internet or library resources available to you, undertake research to provide detailed answers to the following questions:

   a) From which countries did many West German guest workers come? It would be helpful to list these countries in order according to how many guest workers they provided.

   b) During which years did this international migration flow take place, and in which year(s) did it peak?

   c) What were the perceived pull and push factors that influenced these migrants' decisions to move?

   d) How were the migrants treated by their hosts (i.e. their German employers and the local people)?

   e) What proportion of guest workers eventually returned to their original countries? What reasons did they give for returning?

2. a) Undertake a survey of adults in your own local area to obtain information about: how long they have lived in their present house, the town/region where they had moved from (if appropriate) and the main reasons for their move to that locality.

   b) Collate and then analyse the data you have collected, using such terms as 'international migration', 'urbanisation' and 'counter-urbanisation'.

Examination-style questions

1. a) Name any two migration theories, name their authors and state the years in which they were first published.

   b) State which of these two theories is, in your own personal opinion, the more realistic in terms of recent migration flows; give detailed reasons for your decision.

2. With reference to case study exemplars:
    a) Suggest reasons why 'retirement migration' is becoming increasingly popular.
    b) Discuss the potential implications of large-scale retirement migration for popular retirement destinations.
3. Assess the validity of the following statement: 'Migration is now the most important component of population increase in many parts of the world'.
4. Use the concept of pull and push factors to explain why depopulation often takes place in rural areas. Base your answer on a range of historical and economic contexts.
5. With reference to named examples describe the effects of both in- and out-migration on the communities involved.

4 Urbanisation and Population Distribution

1 Introduction: Putting People in their Place

This chapter examines the factors that determine population distribution: those factors that influence people where, or where not, to live. It does this by considering population distribution at increasingly larger scales, in this order of magnitude:

- the site and situation locations of individual settlements
- population distributions within urban areas

- national population distributions
- the global population distribution.

Section 7 examines a major factor in changing population distribution – urbanisation – with particular respect to Brazil and its largest city.

2 Site and Situation

All **settlements** – even the great **conurbations** of the world – began as small clusters of buildings. Those clusters with a good location thrived and those locations that proved to be advantageous over many centuries become larger, more important and so progressed upwards through the levels of the settlement hierarchy. If follows that an understanding of what constitutes 'good' and 'bad' sites is essential to a genuine understanding of urban population distribution.

The **site** of a settlement is the place at which it began to exist – the land on which its very first buildings were constructed.

The **situation** of a settlement is its location within the much larger area of land which surrounds it. 'Natural' situation factors within this area include its physical relief and drainage patterns, and the availability of any useful raw materials that occur there. The main 'human' situation factor is the transport network to which a settlement is linked.

The following case study provides details of both the site and the situation of Preston, a city that has developed very successfully due to its excellent location factors.

CASE STUDY: THE SITE AND SITUATION OF PRESTON

Figure 4.1 is a simplified map of the site of Preston, a settlement in central Lancashire whose successful development over many centuries was officially recognised in 2000 by its selection as England's Millennium City. One of the reasons for Preston's success is that its original site has since proved to be extremely well chosen. One of its prime site location factors was being close to the **confluence** of the River Ribble and one of its **tributaries**, ensuring that its people always had a reliable and abundant supply of water for drinking, washing and industrial purposes. Preston's site is on higher ground overlooking the river valley below, which means that it has never been in danger of being flooded when the river overflows onto its **flood plain**. The steep valley side between river and town also gave some measure of defensive protection against invaders from the Irish Sea. Wood needed for heating, cooking and building construction was

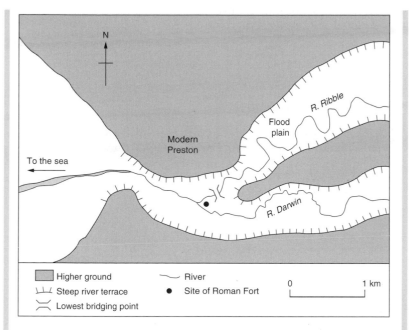

Figure 4.1 The site of Preston

readily available to the north of the settlement at Fulwood, now one of Preston's largest and most affluent suburbs.

Preston also had the advantage of been well situated. Preston's advantageous situation (Figure 4.2) included being very near to the Ribble's lowest bridging point, establishing it as one of the key transport nodes within north-west England's road and railway networks. Its location at the head of the Ribble estuary gave it the potential to develop as a port. This combination of excellent site and situation locations has allowed Preston to become **multifunctional** and adapt successfully to changing situations. The most important of Preston's past and present functions are listed below:

- Military function: the strategic value of its site was instantly recognised by the Romans, who built a fort near to its lowest bridging point. The army's presence in the city is still maintained at Fulwood Barracks, in north Preston.
- Residential function: Preston's population now exceeds 140,000.
- Industrial function: the Industrial Revolution prompted Preston's rapid expansion as one of the main weaving towns in Lancashire's booming cotton textile industry. The town's population increased from 11,887 to 24,575 between 1801 and 1831: a remarkable rise of 107%. Modern Preston has a much wider range of industries, including aircraft manufacture.

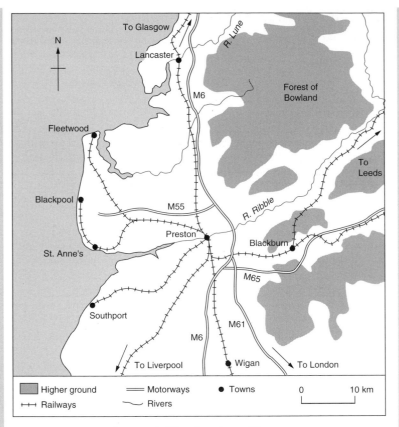

Figure 4.2 The situation of Preston

- Maritime function: Preston was an active port for many years, until its large enclosed dock (Europe's largest single dock basin) was adapted to use as a yacht marina in the 1980s.
- Transport network function: Preston's importance as a road and railway centre has already been noted, and was again emphasised by the construction of the Preston bypass, the UK's first stretch of motorway, opened in December 1958.
- Service function: Preston provides a wide range of retail and other services for the central Lancashire region. Its recreational and entertainment facilities include The Guildhall and Preston North End Football Club.
- Administrative function: Preston is now the 'county town' of Lancashire, a role it assumed from Lancaster because of its more central location within the county.

3 The Settlement Hierarchy

This section of the chapter investigates the size of settlements, which is a particularly dynamic aspect of population geography because of the rapid changes taking place within it. One example that illustrates the speed of this change is the size of London's population compared to those of the world's other major cities.

In 1801 London had a population of 1,114,644 and was widely acknowledged to be the largest city in the world. Now, even though Greater London and its adjacent urban areas have a combined total population of over 12 million people, the UK capital has slipped to 19th place in the global 'league table' of populations of **metropolitan regions** (now often also referred to as **agglomerations**). It seems, therefore, that there is an irresistible tendency for the largest urban areas to grow at an ever-increasing rate, hence the introduction of terms such as conurbation, **megacity**, **megalopolis** and **millionaire city** to describe those places that are at the pinnacle of the **settlement hierarchy** (Figure 4.3). This is a system by which settlements are arranged according to their population size, the number and range of their services, and the direct distances between them. The hierarchy attempts to differentiate between settlements within rural and urban areas, even though this cannot be done with a high degree of precision because national definitions of what constitutes an urban area vary so much (Figure 4.4).

While the nature of each of the hierarchy's components is summarised below, it is important to be aware of the absence of any universally agreed definitions for these components. For example, each country has evolved its own requirements for city status and there is often a degree of confusion even within countries. This is certainly true of the UK, where the custom has been for any settlement having a cathedral (the headquarters church of a bishop's diocese) to be accorded city status automatically. The classic example of this is the

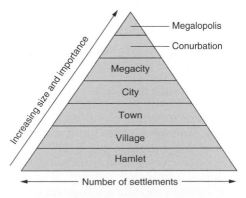

Figure 4.3 The settlement hierarchy

Minimum settlement population required for urban status	Country (ies)
30,000	Japan
10,000	Switzerland
2,500	USA
2,000	France
1,000	Australia, Canada
200	Denmark, Sweden

Figure 4.4 Variations in national requirements for urban status

village-city of St David's in the far south-west of Wales, which is a 'cathedral city' even though its population is only 1600.

- Megalopolis: this term is reserved for the largest of all urban areas. It was first used in 1961 by Jean Gottman to describe what was originally nicknamed 'Bosnywash', but which is now usually short-ened to 'Boswash', the huge urban sprawl in the north-east of the USA extending southwards from Boston, through New York, Philadelphia and Baltimore to its tip at Washington (total popu-lation 41.35 million). The world's other three examples of mega-lopolises are:
  - 'Sansan', San Francisco–Los Angeles–San Diego, in California (total population 27.8 million)
  - Tokyo–Yokohama–Nagoya–Kyoto–Osaka–Kobe, on the east coast of the Japanese island of Honshu (total population 58.65 million)
  - the 'Ruhr Coalfield' industrial region of Germany, the only mega-lopolis to be located totally inland and having a decreasing popu-lation at the present time. This very densely populated region includes Bochum, Cologne, Dortmund, Duisburg, Düsseldorf and Essen (total population 5.8 million).
- Conurbation: a very large urban area which includes at least one major city and often has numerous towns within its metropolitan region.
- Megacity: a super-city having a population of at least five million people.
- Millionaire city: a city having at least one million people – *not* a city totally inhabited by the rich! The world's first city to achieve mil-lionaire status was Beijing, in the late 1700s.
- City: any very large built-up area having a wide range of adminis-tration, commercial, educational, industrial and transportational functions at a regional scale – possibly including airport facilities.
- Large town: an almost completely self-sufficient urban community having a range of facilities including a hypermarket, a hospital and a number of secondary schools.

- Small town: a settlement which is able to fulfil many of the basic requirements of its own inhabitants as well as those in the rural area surrounding it, e.g. a doctor's surgery and a small secondary school.
- Village: a small settlement that may include some basic facilities such as a post office, several shops offering a range of lower-order goods required daily, a public house, a garage and possibly a primary school.
- **Hamlet**: a cluster of houses with very few, if any, services. Many hamlets have a public roadside telephone box, but most are too small to justify having their own church.
- Isolated building: remote building such as a farm, public house and castle.

4 The Central Place Theory

Most settlements provide essential services not only for their own inhabitants but also those people who live in their surrounding **spheres of influence** (also known as **catchment areas**). The larger a settlement is, the greater its sphere of influence is likely to be. One way of displaying such a settlement-dependency hierarchy is shown in the highly idealised form in Figure 4.5 – the so-called **Central Place Theory** devised by Walter Christaller in 1933. This was based on observations that he made in southern Germany which led him to believe that small settlements were much more numerous than larger ones and that central places of the same size tended to have very similar ranges of facilities. He also utilised two concepts that are still relevant to the study of settlement location and growth:

- **range**: the maximum distance people are willing to travel to obtain a particular service
- **threshold**: the minimum area required to provide enough potential customers to make a service viable.

However, Christaller's pioneer work is now believed to have a number of serious weaknesses. Most of this criticism concerns the following assumptions that he made at the time:

- That transport costs would increase at the same rate in all directions from every settlement. This is a situation that could occur theoretically, but only on a completely flat and totally uniform land area (an **isotropic plain**).
- That the human population would be evenly distributed.
- That customers act in a totally rational way and therefore always opt to travel to the nearest central place in order to obtain the goods and services they need. This is clearly not the case, because people have highly individual preferences and often regard travelling distance

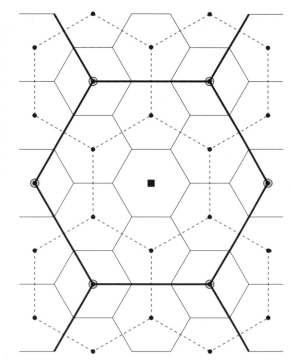

Figure 4.5 Central Place Theory

as far less important than other factors such as travelling *time*, parking facilities, and the locations of family and friends.

- That all settlements of the same size provide a similar range of functions. This is not true, because many settlements have become highly specialised in functions such as tourism and manufacturing industry.

- That 'chance' is not a factor in the establishment and subsequent development of individual settlements. There are many examples which disprove this assumption, e.g. the establishment of the steelworks at Shotton, in north Wales, which was selected as the preferred site from a range of equally attractive alternatives simply because that was where his future wife lived!

- That all customers have the same income and hence identical spending power. This is clearly not the case; employed people earn different rates of pay according to the nature of their work; some people are unemployed or retired; some individuals are very careful with their money, whereas others may be spendthrifts.

- That it was not necessary for the central place concept to be flexible enough to allow for changing circumstances. Settlements can only survive at their present size or expand further if they are able to adapt to changing and challenging circumstances.

5 Population Distribution Within Urban Areas

Every settlement is different for a wide variety of reasons. The site may physically limit the direction(s) in which a settlement can grow and can also restrict its size. For example, towns built in narrow valleys usually become **linear** (long, but narrow) in shape as in the case of the former coalmining towns in the South Wales valleys. Many coastal holidays resorts are also linear, because their main business and holiday accommodation zones have to be as near to the seashore as possible. However, the morphology of any town is determined to some extent by the distances its residents have to travel to shop and possibly work in the CBD. This is the chief reason why most settlements tend to be **nucleated**, which means that they are much more compact and generally rounded in shape. The layout characteristics of urban settlements is called **urban morphology**, a topic now examined in some detail because it helps us to understand the reasons for varying population distribution and densities within built-up areas.

The internal layout of a settlement can often tell us a great deal about its past history. Some settlements are very old and the basic street patterns of their central areas may not have changed for hundreds of years. Many British settlements expanded rapidly during the Industrial Revolution, at a time when most housing was terraced and built along streets that followed a rigid grid-pattern often adjacent to the mills and factories that provided work for their occupants. Some settlements are almost entirely modern creations, and their road and land-use patterns reflect the changing demands of urban planning controls, restrictions that only became a standard feature of life in Britain during the twentieth century.

One very effective way to study the layout of a settlement is to identify its component **urban zones**. A zone is a part of an urban area that has a particular function and differs significantly from any of its adjacent zones. It is quite common for large residential areas to be divided into separate, very distinctive zones because their street layouts, housing types and other key characteristics, such as the provision of recreational open space, are so different.

Figure 4.6 shows a type of pattern of urban zones that is typical of many long-established British towns and cities. This particular pattern was devised in 1924 by a US sociologist called E.W. Burgess and is usually referred to as the **Concentric Urban Model** because it consists of a series of different-sized circles all sharing the same centre. Like most models, this urban zone pattern is very simplified and not intended to be a perfect reproduction of any particular real-life settlements. In fact, the concentric ring model is so idealised that it is only likely to reflect situations within a totally uniform physical environment, one in which urban expansion can take place unhindered in every direction due to the total absence of constraints such as coastlines, rivers, lakes, marshes and steep gradients. The real value

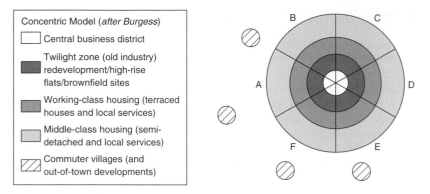

Figure 4.6 Concentric Model of urban zones

of this model is in providing a tool by which settlements can be analysed and then understood more easily. Each of the major urban zones in Burgess's model is described separately below, with particular regard to their population density characteristics.

a) The central business district

The **central business district (CBD)** is the zone at the heart of a settlement and provides most of the services needed by its people. It is the chief commercial and administrative district, and so attracts many daytime visitors, most of them shoppers or workers commuting daily into the zone. Very few people can afford, or indeed may wish, to live in the CBD zone. There is usually very little recreational open space, traffic congestion is a constant difficulty, and high air and noise pollution levels pose serious environmental problems for residents.

CBD land and property values are so high that they are well beyond the financial means of most people. Those people who do live there are either the super-rich who can afford penthouse suites or business people (e.g. public house and hotel staff) who occupy on-site accommodation. For example, the City of London had well over 100,000 inhabitants in 1801, but only 6500 in 2001. Its population declined as its residential areas were displaced by more profitable, and therefore much more expensive, commercial properties. Most people now commute to this central area instead of living in it.

b) The inner residential zone

This type of zone mostly dates back to the nineteenth and early twentieth centuries, and consisted originally of a mixture of industrial, commercial and residential properties. The housing typical of that period was terraced, in order to achieve maximum population densities, and one of its earliest forms was a back-to-back arrangement. Public transport at the beginning of that period was still in its infancy and working hours were very long, so most employees had to live within a short walking distance of their place of employment. These areas have since proved especially popular with ethnic communities and young families because of lower house prices and accessibility to employment opportunities.

During the 1960s and 1970s, many of these areas were subjected to drastic **urban renewal** programmes that reduced their population densities so much so that they also became known as **zones in transition** and **shatter belts**. Thousands of the smaller terraced properties were replaced by high-rise blocks of flats, but some of the larger houses were bought and expensively renovated to a very high standard, a form of urban renewal called **gentrification**. Within London, parts of Islington and Notting Hill became gentrified in the 1970s and areas within Wandsworth became gentrified during the 1980s when young, highly-paid professionals began to appreciate the advantages of living within a short travelling distance of their inner-city work places as well as having easy access to central London's vibrant social and cultural life.

The case study which follows shows how the Toxteth area of central Liverpool was **redeveloped** and **renovated** during the 1980s with a consequent reduction in its resident population. Some inner residential zones have been almost completely adapted to university student accommodation, which means that their populations are not only transient, but may be virtually non-existent during extended holiday periods within the academic year.

c) The suburbs

Towns expanded very rapidly during the 1920s and 1930s due to technological developments in public transport such as electric-powered trams, diesel-engine buses and improved railway networks. The Underground, for example, was the chief stimulus for many inter-war, house-building programmes in London's outer **suburbs**. The growth of private car ownership, especially in the 1960s and 1970s, was also a key factor in the increased rate of **urban sprawl** (outward growth) that took place and accelerated the trend towards longer commuting journeys between home and work.

Most private residential developments took the form of two-storey, semi-detached housing. Local authorities also became very active in building housing estates of rented, council-owned accommodation. Both private and public housing developments at that time resulted in much lower population densities within the suburbs than in the inner residential areas.

d) Industrial zones

Industry occupies much less urban land than housing, but it too can occur in almost any part of a built-up area. Industrial zones first developed during the Industrial Revolution, were usually adjacent to canals and railways, and incorporated clusters of cheap housing for their workers. The rapid expansion of motor transport in the twentieth century made access to main roads a much more important factor and was one reason for the creation of another urban morphology model – the **Sector Model** (Figure 4.7). This model, devised by Homer Hoyt, was based on research conducted in 142 US cities in 1939. Hoyt's model acknowledged that industry is one of a number of land uses that tend to expand outwards from settlement centre in a linear rather than a concentric ring pattern. A common example of this tendency is that high-status housing on the edge of a CBD tends to attract similar housing developments further away from it. In British towns, this often occurred to the south-west of the CBD and any nearby industrial zones, in the certain knowledge that the prevailing south-westerly winds would disperse polluted air towards the less affluent residential zones in the north and east! During the twentieth century, the concept of **industrial estates** became so popular that all major urban areas now have areas devoted exclusively to commercial and industrial uses. These may be on **brownfield sites** within the established urban area, in sectors along routeways and on **greenfield sites** on the rural–urban fringe.

A third urban morphology model recognises the fact that very large settlements develop **secondary business districts**. These tend to be some distance from the CBD, far enough away for them to be profitable suppliers of all but the highest-order goods such as expensive jewellery and haute-couture fashion clothes; these continue to be sold exclusively in the central area. This model, the **Multiple Nuclei Model** shown in Figure 4.8, was devised by C.D. Harris and E.L. Ullmann in 1945, in recognition of this tendency for large cities to

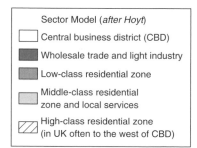

Sector Model (*after Hoyt*)

☐ Central business district (CBD)

■ Wholesale trade and light industry

■ Low-class residential zone

Middle-class residential
zone and local services

High-class residential zone
(in UK often to the west of CBD)

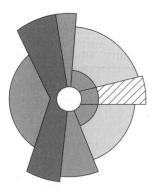

Figure 4.7 Sector Model of urban zones

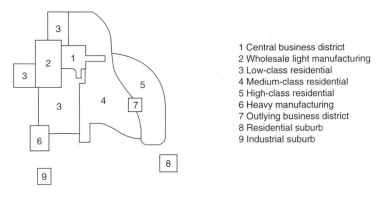

1 Central business district
2 Wholesale light manufacturing
3 Low-class residential
4 Medium-class residential
5 High-class residential
6 Heavy manufacturing
7 Outlying business district
8 Residential suburb
9 Industrial suburb

Figure 4.8 Multiple Nucleii Model of urban zones

spawn suburban, secondary business districts. It is very common for such districts to have developed out of long-established villages and small-town centres that later became part of the city when its outwards sprawl surrounded them. This third model also recognises the modern practice of locating new industrial zones next to secondary business districts, university campuses and airports beyond the rural–urban fringe.

CASE STUDY: URBAN RENEWAL IN TOXTETH

Toxteth is one of Liverpool's inner residential zones. It lies between the city's CBD and the areas of more recent and much more spacious housing around Princes Park and Sefton Park to the south-east of the centre. Toxteth is a zone of great contrasts, partly for historical reasons, but also because of the urban renewal that transformed it during the 1980s and 1990s. In 1981, it was the scene of some of Britain's worst street riots in recent times.

The Toxteth area was first developed during the nineteenth century, when Liverpool became a major global port and its increased employment opportunities led to substantial in-migration and an urgent need for new housing of all types. A class of wealthy merchants was created by the city's shipping and related onshore industrial activities, including the notorious transatlantic Slave Trade. These Liverpool merchants built imposing detached villas and high-status terraces along a new, tree-lined avenue called Princes Road. The thousands of ordinary working class people were housed in hastily built, much smaller terraces on parallel side-streets leading off Princes Road.

As Liverpool declined as a port and manufacturing centre during the second half of the last century, its unemployment difficulties increased. Toxteth was one of the inner residential zones most seriously affected by the city's decline and received considerable financial support through government grants.

The riots that took place there in 1981 were so serious that the government re-allocated funds intended for its new towns programme into inter-city renewal. Urban renewal in Toxteth took two forms: high-status housing that was still structurally sound was renovated and given a new lease of life; the workers' terraces were demolished and replaced by low-density modern housing including pleasant cul-de-sacs and clusters of small bungalows for the elderly built under the redevelopment programmes. Some of the land on which demolished housing once stood was devoted to new sports facilities and landscaped gardens, which greatly improved the quality of life of local residents. As a result of all these measures, much of Toxteth has been completely transformed and its overall population density substantially reduced, while still allowing its rich ethnic diversity to be retained.

CASE STUDY: POPULATION SEGREGATION IN BELFAST

Belfast, the primate city of Northern Ireland, has been called a divided city in recent years. This is because many of its residential zones are occupied mainly either by Catholics or Protestants, their segregation having taken place for both religious and political reasons. This situation is in direct contrast to that in other British cities, in which people usually choose where to live on economic, social, but, increasingly, ethnic grounds.

Early maps of Belfast show that some segregation was taking place as early as 1685. Belfast had already become the chief port for trade with England and Scotland, and this encouraged people from both of these Protestant countries to seek work there. Many of these Protestant immigrants worked in the docks and mills that wove flax into high-quality linen cloth. Later, Belfast became an important centre for shipbuilding and engineering, and this accelerated the immigration rate from Britain. Transmigration within Northern Ireland also boosted Belfast's development as thousands of Irish Catholic farmers abandoned their land in search of better-paid work in the city.

Unrest between the two religious communities grew steadily worse and even the partition (political subdivision) of Ireland into two separate countries in 1922 failed to eliminate the dangerously high level of tension between Catholics and Protestants. The newly created republic in the south of the island (then named Eire) refused to accept that the province of Ulster (the much smaller north) should remain part of the UK. The inter-island and rural–urban migrations into Belfast put great pressure on the city's resources. Poverty, overcrowding and widespread unemployment were the result, and these led to increasing jealousy and bitterness between the two opposing communities.

The most recent phase of violence (called 'The Troubles') began in 1969. During the 30 years of partisan violence that followed, rival mobs clashed openly on the streets, property was looted, banks were raided and the owners of small businesses were blackmailed into handing over protection money. Murders and muggings became commonplace, one of the most unpleasant events being a bomb attack by one of the Province's paramilitary groups on a Remembrance Day service at a war memorial.

Unemployment has certainly helped to heighten inter-community tensions in recent years and Northern Ireland has consistently experienced one of the highest rates of unemployment in the UK's Government Office Regions (GORs) (Figure 4.9). This is due to the contraction of the city's main industries. The workforce of shipbuilders Harland & Wolff decreased from over 25,000 in 1970 to fewer than 5000 today. The linen textile industry has virtually collapsed due to competition from cheaper artificial fibres such as nylon and rayon.

It is hardly surprising that both Catholics and Protestants cluster closely together in their 'own' residential safe-havens. Thousands of families have migrated within the city to create the highly segregated population distribution pattern shown in Figure 4.10. Internal migration on this scale has been possible because large areas of terraced housing have been redeveloped and replaced by new estates on the rural–urban fringe; one such suburb is the Catholic 'stronghold' of Andersonstown. The inner-city zones also have similar patterns of segregation. Two streets that link the inner and outer residential areas and became known worldwide are the Falls and Shankhill Roads, between which a so-called Peace Line was created – a high concrete dividing wall much defaced by sectarian graffiti.

Discussions between the UK and Irish governments have resulted in an improved situated called 'The Peace Process' and the signing of the 'Good Friday Agreement' on 10 April 1998, after which fewer bombings and murders have taken place. However, the continued frequency of vicious muggings and beatings with homemade weapons such as clubs embedded with long sharp nails sadly reinforces the on-going need for segregation

Figure 4.9 UK government office regions

within the sectarian communities of Belfast, Londonderry and Northern Ireland's other urban communities. Angry scenes between parents outside a primary school in September 2001 indicate that it is likely to be many decades – and possible many more generations – before the endemic bitterness within Northern Ireland really begins to fade.

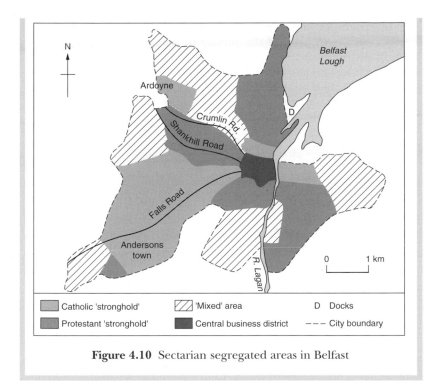

Figure 4.10 Sectarian segregated areas in Belfast

6 National Population Distribution

The following two case studies provide ample evidence that population distributions can vary widely between individual countries.

CASE STUDY: POPULATION DISTRIBUTION IN BRITAIN

Britain's population distribution is very variable and has changed significantly over the past two millennia. Events of particular significance in this changing distribution include the end of the Ice Age, the Roman and other numerous invasions, the Industrial Revolution and increasing government intervention during the second half of the twentieth century. The earliest distribution patterns were due chiefly to climatic factors, the morphology (physical relief) of the landscape, and changing economic, political and technological factors.

International migration and transmigration have both been major factors in most of the phases of Britain's changing population distribution.

Figure 4.11(a) shows the impact of the major events within pre-industrial Britain on its population distribution. Inevitably, the Ice Age restricted any human inhabitation to the extreme south of England – beyond the maximum extent of the ice sheets that had originated in Scandinavia and the regions of much higher land in northern Britain. The period of global warming that triggered the end of the Ice Age led to higher sea levels and created the English Channel by flooding a long depression in the land between continental Europe and what became the British mainland. As the ice sheets melted, the land over which they had retreated was covered by **tundra** similar to that in the present Arctic regions of North America and Asian Siberia. Over a long period of time, this inhospitable environment was succeeded first by coniferous forests and then deciduous woodland. People began to give up their traditional nomadic life-style and created permanent settlements on the coasts and less densely vegetated uplands.

After their successful invasion in 43 CE, the Romans established their first commercial centre at a safe bridging point across the River Thames, and in doing so began the two millennia history of Britain's largest and, for most of that time, its capital city. In fact, nearly all the Roman towns were built in lowland areas, settlements in higher land being restricted to isolated military forts. History repeated itself during the fifth to tenth centuries when a series of invasions by Angles, Saxons and then Vikings took place. The Norman invasion of 1066 merely consolidated the existing population and settlement distribution patterns, which remained basically unchanged until the Industrial Revolution of 1760–1850. Prior to this, about 80% of people lived in rural areas; after it, about 80% lived in urban areas.

Figure 4.11(b) shows the *post*-Industrial Revolution population pattern, indicating that most people are still living in the lower areas. It also shows that there are a number of dense clusters that, at first glance, do not appear to follow any particular distribution pattern. The location factor shared by all these conurbation-size population clusters – except London – was the existence of extensive coal seams. In some areas, the coal measures included deposits of **blackband iron ore** that, together with local limestone and plentiful water, provided all the raw materials needed to produce iron and steel. Most of these industrial conurbations specialised in manufacturing certain types of goods. For example, Lancashire's thousands of mills spun and wove cotton textiles, whereas those in west Yorkshire, only a short distance away on the eastern side of the Pennines, became expert

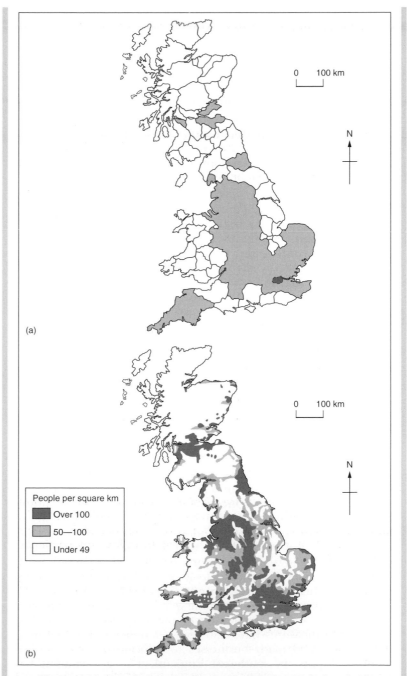

Figure 4.11 Population distribution in Britain: (a) before and (b) after the Industrial Revolution

in the production of woollen goods. The West Midlands quickly earned a worldwide reputation for precision engineering and even boasted that it could produce anything from a steel pin to a complete train.

Figure 4.12 locates Britain's chief coalfields and identifies the manufactured goods in which each coalfield specialised. In spite of not being located on a coalfield, London developed the broadest industrial base of all Britain's conurbations, producing many of the goods needed by the domestic Greater London market. London even had a number of shipyards for a period of time, until the need for ever-larger ships made their closure inevitable. It was also the site chosen for Ford's largest car assembly plant in Britain.

After the First World War, some of the major traditional coalfield-based industries entered a period of almost terminal decline. Britain had permanently lost many of its overseas markets for exported coal during the five years of conflict, and cheaper imported goods forced many factories and textile mills to close. Much obsolete machinery in these regions was scrapped and not replaced. It was in the London–Birmingham axis that most investment in new manufacturing plant was made.

The period of greatest economic hardship for the traditional industrial areas occurred during the 1920s and 1930s, in the period now referred to as the Great Depression. As unemployment rates increased in these areas of industrial decline, the quality of life of their people decreased to such an extent that political pressure forced the government to take remedial action. Later governments have felt obliged to support the weaker economic regions ever since. In 1966, the government created the first **development areas** to help those regions experiencing chronic unemployment (Figure 4.13). **Intermediate areas** were created four years later for those regions requiring less support because their unemployment-related problems were not quite so serious. A wide range of **incentives** was offered to expanding companies to encourage them to invest in these supported areas. Such pull factor measures included tax relief, financial grants towards the cost of new buildings and equipment, and subsidised training and removal expenses for key workers.

The government's push factors included requiring companies to obtain **Industrial Development Certificates** prior to building new factories, certificates that were deliberately made very difficult to obtain for any new developments outside the assisted areas. Such measures have had some beneficial effect because they encouraged many businesses to relocate outside employment 'hot-spots' in south-east England. The government has itself relocated many of its own offices outside the south-east and

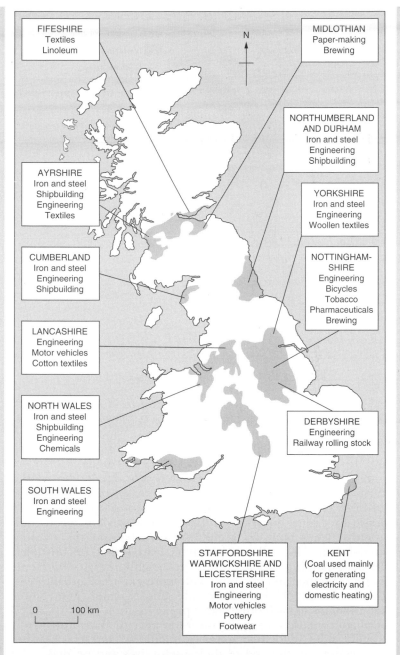

Figure 4.12 British coalfields and their traditional industries

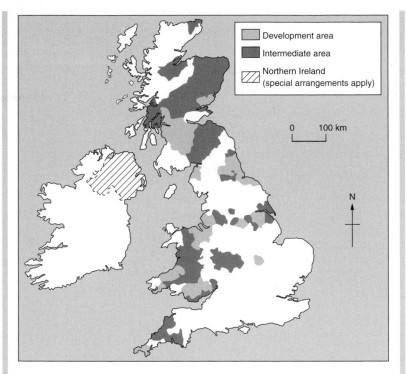

Figure 4.13 UK development and assisted areas

its incentives were largely responsible for the successful high-tech industrial developments in the Scottish lowlands around Glasgow, the M4 'corridor' between London and south Wales, and the university town of Cambridge in East Anglia.

Britain's towns and cities expanded very rapidly during the 1920s and 1930s, and there were several reasons for this accelerated rate of urban sprawl. Increased car ownership and people's desire for larger houses with garages and gardens led to new suburbs being built on greenfield sites beyond the existing rural–urban fringe. Town planners feared that uncontrolled urban expansion would result in neighbouring settlements joining together and creating even more extensive urban areas. The nearness of the Merseyside and Greater Manchester conurbations to each other was a prime example of what caused so much concern during the inter-war period. Many nineteenth century inner residential zones of traditional, terraced housing were coming to the end of their useful lives and needed to be replaced. The preferred method of urban renewal at that time was (comprehensive) redevelopment, the demolition and

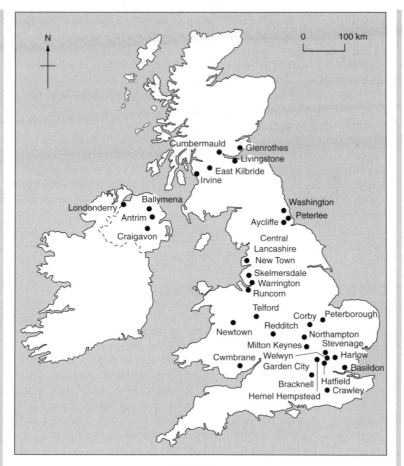

Figure 4.14 UK new towns

complete replacement of existing properties. This process resulted in much lower population densities within the inner residential zones and led to an urgent need to re-house the **overspill population**, those people who were made homeless by this form of urban renewal. Constantly building on urban fringes to provide new accommodation for these people would, of course, merely add to the urban sprawl the planners wished to avoid.

The planners' solution to this dilemma was to build a series of **new towns**, many of which were really **expanded towns** because they were built next to existing rural settlements (Figure 4.14). The New Towns Act of 1946 led to the building of Stevenage and five similar towns around the London conurbation, but well outside its **green belt** of protected land which had been established to counter the capital city's tendency to sprawl into its surrounding countryside. Similar clusters of new towns were approved for all the other major UK conurbations so that their overspill populations would not have to migrate to distant and unfamiliar parts of the country. Many of these so-called new towns are now over 50 years old and most accommodate many more people than their intended population of about 50,000, a figure then accepted as the minimum for an urban community to be fully self-contained in terms of employment opportunities, service provision, and the availability of social and recreational facilities.

The new towns initiative created new accommodation for about a million people and in doing so provided one of the outstanding examples of twentieth century governmental impact on population distribution.

CASE STUDY: POPULATION DISTRIBUTION IN BRAZIL

Brazil, South America's largest and most heavily populated country, also has a variable population density pattern. Figure 4.15 shows that only a relatively small proportion of the total land area is densely populated. The coastal region of the south-east, which is especially heavily populated, includes the country's three largest urban settlements: São Paulo, Rio de Janeiro and Belo Horizonte. In fact, over 90% of all Brazilians live on or close to the Atlantic coastline. Safe, natural harbours have encouraged transatlantic trade with Europe for centuries – especially with Portugal – and Brazil is unique in South America for having Portuguese, not Spanish, as its official language. This language divergence results from an agreement between Spain and Portugal, which established the Line of Tordesillas in 1494 and effectively divided the South American continent into two separate areas of European colonial influence. This was done to reduce the risk of these two colonial powers entering into conflicts that would have harmed the trading interests and the economies of both countries.

The coastal south-east has plentiful and reliable natural water supplies, and a combination of high temperatures and well-distributed annual precipitation has enabled its farmers to achieve high arable yields. Coffee thrives in this region because its chief growing requirements are met: an average monthly temperature within the range 21–26°C, an absence of overnight frosts, an evenly distributed rainfall of 1800 mm and deep, well-

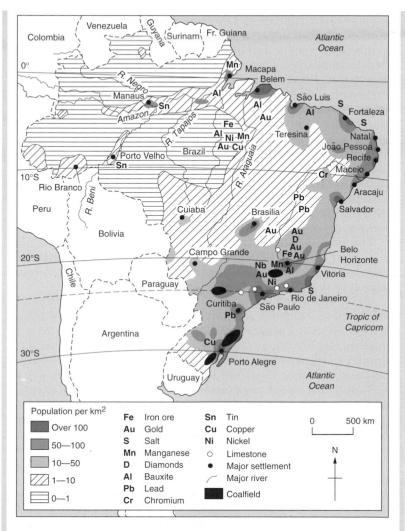

Figure 4.15 Brazil: population distribution and natural resources

drained fertile soils. The **terra rossa soils** of volcanic origin that occur on the coastal plateau inland from São Paulo are ideal for this commercial crop. Tobacco, cotton and vines are other important regional cash crops. Figure 4.15 also locates those valuable raw materials that stimulated industrial development in the southeast.

Locally produced gold was a key export to Europe during the eighteenth century, and was a major factor in the early prosperity and growth of this region. The numerous coalfields and extensive iron ore deposits inland from the coast were especially important in this development and formed the basis for its steel and motor vehicle industries. One of the largest integrated steel-making plants is at Volta Redonda and Brazil's car industry began in 1919, when 'Model T' Ford cars were first assembled locally from imported parts. Today, Ford is a major employer in the Belo Horizonte region and many **transnational corporations** (TNCs) with headquarters in the USA, Germany Italy and Japan have branch factories there. The Fiat car assembly plant, for example, has been in operation since 1978 and now employs well over 10,000 workers. Both the steel and the motor vehicle industries have had a significant **multiplier effect** within south-east Brazil, each having stimulated the growth of many product-dependent component manufacturing companies.

The steel industry was crucial in the development of Brazil's important shipbuilding industry. Offshore petroleum deposits have led to a **petrochemical industry** that serves the domestic market of over 160 million Brazilians.

Other major, but much more remote, settlements are Brasilia, Manaus and a series of isolated ports on the most easterly coast, notably Belem, Fortaleza, Recife and Salvador, which benefit from the excellent network of railways serving the coastal region. Brasilia was created in 1960, to replace Rio de Janeiro as the capital city (hence its name) and to attract industry, commerce and especially people away from the densely populated coastal regions.

Manaus is the most remote of all Brazilian cities. It is situated 1000 km inland, at the confluence of the Amazon and Negro rivers, and is the capital of the **Selva** equatorial forest region. The city has good road access to the 6400-km long Transamazonica Highway (built in the 1970s), which links the western and eastern extremities of the country. It was hoped that this highway would stimulate transmigration from the overcrowded south-east, but many of the early government-sponsored colonists found that the dense natural vegetation was difficult to clear and the soil quickly lost its fertility due to leaching and lack of decomposing leaf litter which followed tree-felling. Figure 4.16 illustrates the nutrient cycle which is a basic component of the equatorial rain forest **biome**.

In pre-colonial times, native Indians such as the Yanomami were able to survive due to their sustainable – not intensive – land-use strategies within their own large tribal territories. Recent developments such as **open-cast** metal ore extraction and extensive cattle ranching have attracted some people to the deforested areas, but have not resulted in greatly increased population

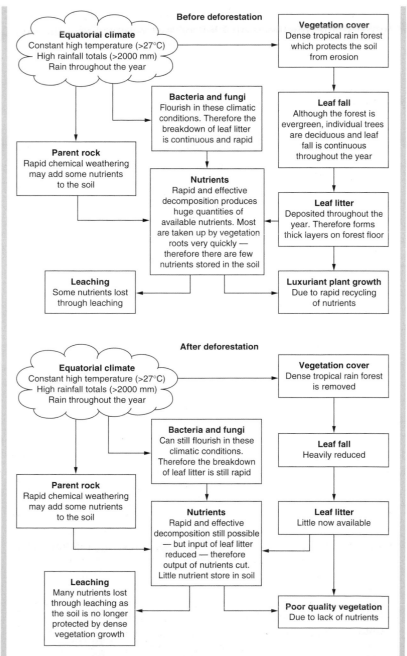

Figure 4.16 Tropical rainforest nutrient cycle: before and after deforestation

densities. The most dramatic of all these recent developments followed the discovery in 1967 of vast reserves of iron ore, copper, bauxite and nickel within the Carajas Mountains, deep inside the Selva. This development stimulated major engineering projects such as railway and hydroelectric power dam construction but, as is usually the case with such projects, it was only its initial construction phase which was labour-intensive – the long-term workforce being much smaller.

The north-east of Brazil has some areas of profitable agriculture, but these are restricted to the coastal strip and along the banks of the major rivers such as the São Fransisco. The main reason for this is climatic, this region having a variable rainfall distribution pattern in which precipitation during the months of July to December in most years is negligible. This pattern makes the region extremely vulnerable to periodic droughts, some of which have lasted five years and left millions of people destitute. The natural vegetation is a form of semi-desert scrub vegetation called **sertao** and subsistence farming is the dominant form of small-scale agriculture.

Plantation-grown sugar cane was a major product for many years, but has been in decline for some time. The region has relatively few valuable industrial minerals to compensate for these problems and its economic decline has resulted in substantial transmigration to the much more prosperous south-east.

7 Urbanisation

Urbanisation within LEDCs has been one of the outstanding demographic trends over the last half century. Brazil's rates of urbanisation (Figure 4.17) have been typical of most LEDCs and resulted in the challenges described in the following case study.

Year	Urban population in Brazil (%)
1940	31
1950	36
1960	44
1970	55
1980	77
1990	78
2000	80

Figure 4.17 Brazil: changes in percentage urban population, 1940–2000

CASE STUDY: THE IMPACT OF URBANISATION ON SÃO PAULO

Brazil's rapid urbanisation since the 1960s has been due mainly to rural–urban migration from its drought-prone north-east to the cities of the industrialised Atlantic Ocean coastal strip further south. The south-east proved particularly attractive because its industrial developments created new employment opportunities. As a direct result of this high rate of transmigration, São Paulo's population rose from 3.6 million in 1960 to 9.6 million in 1991 (an overall annual increase of 5.4%, although the rate in some residential zones was more than double this figure). Even the 45-year-old capital city of Brasilia, 800 km from the nearest stretch of coastline, has experienced widespread *favela* development similar to that shown by Figure 4.18 on page 132.

The majority of these Brazilian transmigrants were so poor that they could not afford to enter the formal housing market. Their only option was to build flimsy shelters from whatever scrap materials they could find, usually illegally on land belonging to the city council or private individuals and companies. These primitive shelters quickly formed dense clusters, most of them located on the outer edge of the rural–urban fringe, causing the city to sprawl constantly outwards. Others developed on smaller sites close to the CBD – some of them on steep, marshy or polluted land that had been left as undeveloped wasteland, but are conveniently situated within a short walking distance of city centre jobs. The generic term for these clusters of spontaneous, illegal dwellings is **shanty towns**, but the Brazilians' own name for them is *favelas*; equivalent terms are *barriadas*, *bidonvilles* and *bustees* in Latin America, north Africa and India, respectively.

Most *favelas* have no electricity or plumbing. Mud paths with open flowing sewers beside them separate the shelters, adding greatly to the health risks to which all deprived, densely populated areas are prone. There is little hope of personal advancement for people living in the *favelas* unless they can obtain more than a basic education that will enable them to break the cycle of poverty in which they have become trapped. The *favela* dwellers rarely live under such conditions out of choice; they are not lazy by nature and do as much as they can to improve their quality of life; they are caring parents and as house-proud as it is possible to be under such difficult living conditions. Many improve their shelters by adding a layer of concrete next to the flimsy scrap-material walls to make them more weatherproof. Some even add a second storey, although doing this can be risky as many *favelas* are built on unstable marshland and steep hillsides prone to landslides. The more enterprising members of the community create small businesses to meet some of their *favela*'s needs. The most desperate residents can seek short-term help from one of the many global charities operating in south-east Brazil, some of them targeting the country's seven million 'street children', orphans who exist under the most basic and exposed conditions, and are often regarded as 'disposable' in the drug culture and inter-*favela* gang rivalry that is rife there.

By 2000, the city's housing deficit exceeded one million residential units. Sub-standard dwellings occupied about 70% of the city's urban area, home to about two million people (20% of the total population). Figure 4.18 shows the locations of the chief *favela* zones within Brazilian cities such as São Paulo.

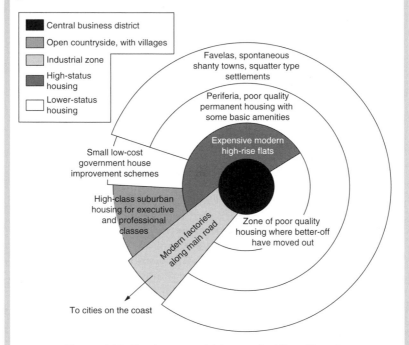

Figure 4.18 *Favela* zones within a typical Brazilian city

a) Self-help projects

A popular and cost-effective strategy has been **self-help projects**, in which the municipal government provides basic services such as electricity, water and sewerage, as well as breeze blocks and other cheap building materials which the local people use to erect more durable homes. Quite often, this work is shared by groups of families who pool their labour and can contribute their own particular individual skills. Although quite basic, such dwellings are a great advance on the unhygienic structures made of highly combustible scrap materials such as wood and cardboard, which were a constant fire hazard.

b) The Cingapura Programme (Projeto Cingapura)

This was a potentially very effective urban renewal strategy because it was based on a highly successful model developed 30 years earlier by the densely populated south-east Asian island-state of Singapore. Its adoption by São Paulo followed the election of Paulo Maluf as mayor in 1992. He believed that his city had consistently failed to address the problem of inadequate housing and that some remedial action was long overdue. The key aspects of the Cingapura Programme are listed below:

- It was a redevelopment rather than a renovation form of urban renewal, targeting slum population within the city's most densely populated zones. A key feature was to provide increased access to basic infrastructure services such as electricity, a reliable water supply and underground sewerage systems. The policy was a 'top-down' programme, totally controlled by the city authorities and involved no local community participation in the consultation and decision-making processes. Eligibility restrictions excluded all single males, persons with a criminal record, old people and even couples who had recently started a family.
- Most apartment blocks were located in areas immediately adjacent to slum housing, so that residents did not have to fund a house move and possibly longer commuting journeys. 'Verticalisation' would make maximum use of the limited area of available land. Early buildings would be low-rise (up to five storeys), with the later phases incorporating 11-storey blocks. Most apartments had two bedrooms, a parlour, kitchen, laundry and bathroom/toilet in 40–50 m² of floor space. Often there were no protective floor coverings such as carpets or tiles and it was common for only two doors to be fitted – one at the entrance and a second for the toilet.
- After completion, apartment ownership passed to the municipal COHAB – the city department responsible for rent collection. A modest fee guaranteed tenants a 25-year period of occupancy. It was theoretically possible for tenants to buy their properties, but most were discouraged from doing so. Despite some modest success, this urban renewal programme was abandoned in early 2001 and only 14,000 of the initially planned 100,000 units were completed.

8 Global Population Distribution

Figure 4.19 shows one way of displaying the distribution of the world's population. It is clear from this map that the population distribution pattern is very irregular, even allowing for the fact that the northern hemisphere has 63% of the total global land area (149 million km²) and is inhabited by about 90% of the total global population (6.5 billion people). The north and south hemispheres' crude population densities are 66 and 15 persons per square kilometre, respectively. This section provides a general description of the global distribution pattern and highlights some of the chief reasons for its many regional variations.

It is also evident from this world map that there are three **primary population concentrations**: Europe, south-east Asia and the east coast of North America. Figure 4.20 displays both the past and current population state of each of the 'populated' continents. The Antarctic landmass does not have a permanent indigenous population, only visiting teams that operate the scientific research bases and provide transport and other basic infrastructure services. The ice 'desert' surrounding the North Pole cannot be included in the number of continents because of its complete lack of land above sea level. However, a very curious situation arose in 2004 when Denmark announced its intention to claim sovereignty over the northern ice cap. Denmark's claim to have ownership rights over this wilderness area is most likely to have been prompted by the increasing accessibility of sub-oceanic deposits of valuable raw materials made possible by the effect of global warming on the Arctic ice cap.

Secondary population concentrations include California, eastern Brazil, south-east Australia, South Africa, stretches of Africa's Mediterranean Sea coast and the northern section of the River Nile valley.

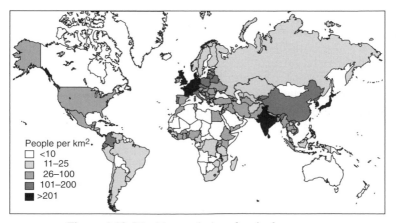

People per km²
- ☐ <10
- ☐ 11–25
- ▨ 26–100
- ▨ 101–200
- ■ >201

Figure 4.19 World population density by country

Continent	Land area (km², millions)	1750	1800	1850	1900	1950	2000
Asia	31.8	63.5	64.9	64.1	57.4	55.7	60.7
Europe	23.0	20.6	20.8	21.9	24.7	21.8	11.8
Africa	30.3	13.4	10.9	8.8	8.1	8.9	12.6
Latin America	20.5	2.0	2.5	3.0	4.5	6.6	8.4
North America	21.5	0.3	0.7	2.1	5.0	6.6	5.0
Oceania	8.5	0.3	0.2	0.2	0.4	0.5	0.5

Figure 4.20 Comparative population change – continent by continent: 1750–2000

Another observation that can be made from these two world maps is that the majority of people inhabit the **temperate** (less extreme) **climate regions**, those lying chiefly between lines of latitude 60°N and 40°S. Two terms that are helpful when contrasting areas of widely differing population density are ecumene and non-ecumene. **Ecumenes** are the inhabited regions; **non-ecumenes** are those with negligible population such as the Sahara Desert, which is literally deserted except for isolated oasis settlements. Non-ecumenes tend to be wilderness areas, remote regions whose ecosystems have not been greatly affected by human activity.

Many factors combine to influence where people do or do not live. Some limiting factors (e.g. international boundaries) are not natural, but most are physical characteristics such as relief, climate and the availability of valuable natural resources. The most important of these location factors are discussed below.

a) Physical relief

Population density generally decreases with increasing **altitude** (the height of the land above sea level). This is because the reduced atmospheric pressure and oxygen content of the air at higher altitudes restricts the body's ability to undertake strenuous work.

These changes may induce 'high-altitude sickness' symptoms such as nosebleeds and severe breathlessness, even though indigenous high-altitude populations do adapt quite naturally over long periods of time to living under such adverse conditions. It is now common practice for international athletes to spend some weeks becoming acclimatised to changes in altitude before actually taking part in competitive events.

Altitude also has a profound influence on climatic conditions. Temperatures decrease at the **adiabatic lapse rate** of about 1°C per 100 m increase in land height, severely restricting the range of commercial crops that can be grown successfully. Most plants are environment-sensitive and have very limited tolerance with respect to growth factors such as temperature, sunlight, rainfall, soil depth and soil fertility. A particularly critical temperature is 6°C, below which plants such as grasses cannot grow.

High-altitude locations often experience strong winds leading to significant temperature reductions due to **wind-chill**. Lower temperatures also modify precipitation patterns, because the **dew-point** (the temperature at which water vapour condenses into liquid form) is reached much more quickly; adverse 'wintry' weather conditions such as mists and blizzards become more common and soils often become less productive due to the leaching of plant nutrients by increased orographic (relief-induced) precipitation. This increased precipitation and the steep gradients on mountainsides accelerate the rate of topsoil erosion, resulting in large expanses of bare, unproductive rock; avalanches and snow drifts make travel extremely hazardous.

b) Climate

Climate is a key factor in population location. Figure 4.21 shows how closely global climatic patterns are linked to latitudinal position. It also helps to explain why the link between latitude and surface temperatures is so strong, by adding two further dimensions to the range of climate factors: the prevailing wind belts and dominant ocean currents, whose effects often combine to have a profound impact on coastal temperature and precipitation patterns. The climatic situation at any particular place is, of course, further complicated by the localised influence of relief, as described above.

In some areas, the temperature of the seawater can have an effect similar to that of latitude. For example, the mid-winter temperatures along the east coast of North America are much lower than those along the Atlantic coast of Western Europe. It is for this reason that some stretches of coastline are actually fringed by deserts (Figure 4.22). The moderating effect of the sea temperature reduces with increasing distance from the coast, with the result that the temperature range between average summer and winter temperatures becomes progressively greater inland. This results in a quite distinct type of **continental climate** occurring (see Figure 4.23, based on seasonal temperature trends across Western Europe). Increasing distance from the sea also affects precipitation because moist on-shore winds invariably become drier as they cross a landmass. This principle of **continentality** is just one of many reasons why very large inland areas tend to be sparsely populated.

c) Natural vegetation

The distribution of the Earth's biomes is shown in Figure 4.24. The grassland, woodland and Mediterranean-type biomes offer the greatest

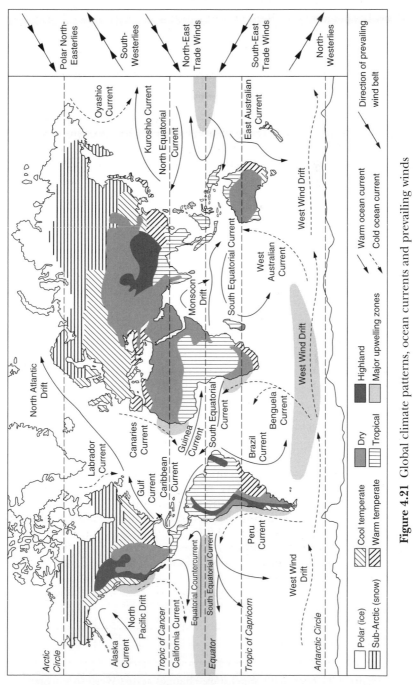

Figure 4.21 Global climate patterns, ocean currents and prevailing winds

Figure 4.22 Juxtaposition of ocean and desert: the Atacama Desert on the Pacific coast of South America

Temperature in °C	Valentia	London	Brussels	Berlin	Warsaw	Minsk	Moscow
January	7	4	3	−1	−3	−10	−15
July	15	17	17	18	19	20	20
Temperature range	8	13	14	19	22	30	35

Figure 4.23 Seasonal temperature changes across Western Europe

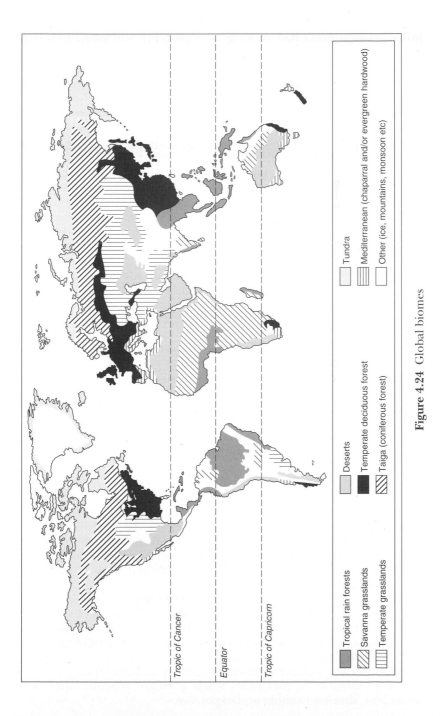

Figure 4.24 Global biomes

potential for agricultural use. The more densely forested regions tend to be negative areas due to their inaccessibility and other inherent problems such as diseases that thrive in humid conditions (e.g. malaria in central Africa). The vegetated tropical regions possess great economical potential, but the frailty of their ecosystems has led to **unsustainable development**, resulting in widespread and often irreversible destruction.

Deforestation of the Amazon Basin and other major concentrations of equatorial rainforest has been undertaken for the sake of short-term economic goals, with the result that permanent population growth within them has been restricted to a few isolated urban centres which process the **primary products** obtained from the deforested areas, e.g. timber, mineral ores and meat. The nutrient cycle on page 129 is a reminder of the ecosystem degradation that occurs when natural vegetation cover is removed without adequate, long-term safeguards such as re-afforestation programmes. Those regions that have a long growing season (because their **temperature range** is 10–30°C and their annual rainfall is evenly distributed) tend to have the greatest agricultural potential of all.

d) Natural resources

Water is the most vital of all natural resources. Under normal temperature and humidity conditions a person may expect to die after 3 days without any fresh water – 10 times faster than someone without food of any kind. Seawater is no substitute for fresh drinking water because the mineral deposits within 'salt' water form a solid layer inside the stomach lining that causes intense pain and can lead to a premature, agonising death. Because of our constant need for drinking water, lakes and rivers have always been prime sites for human settlement. **Artesian basins** provide underground water supplies in many areas of very low total annual precipitation, such as parts of the Australian outback, which would otherwise be uninhabited without expensive piping and irrigation systems. Fresh-water stores such as lakes within the Earth's hydrosphere also have the potential to provide nutritious food. Fish are especially rich in protein, essential for body tissue building and repair, as well as omega-3 fats that can be obtained from most **demersal** (deep-sea) **fish** and helps to maintain the human heart in a healthy condition.

Power can be obtained from the **potential energy** in moving water, using low-technology waterwheels or modern hydroelectric power generating systems. Rivers are capable of eroding and then transporting vast quantities of fine material, much of it later deposited as fertile **alluvium** (silt) on the flood plains of mature river valleys. All water stores are potential natural highways and they proved to be especially valuable in past times when overland transportation was inefficient, time-consuming and expensive.

e) Soils

Soils vary in texture, composition, porosity and agricultural potential. **Permafrost** (permanently frozen subsoils) in high-latitude regions such as Siberia inhibits both arable and pastoral farming.

f) Mineral deposits

Mineral deposits attract economic activity even under some of the most adverse environmental conditions on this planet. For example, oilfields have been developed in many hot desert areas, but are very **capital-intensive** and few workers are needed to maintain them. In contrast, metal-working industries based on mineral ore deposits and coal measures are much more **labour-intensive** and have created some of the largest urban areas in the world.

The case study of population distribution in Britain provides many examples of rapid population growth where the all the raw materials needed to manufacture iron and steel were available, as well as steel-dependent industries such as shipbuilding and vehicle manufacture.

g) Coastal locations

Over half of the world's entire human population lives on land between sea level and an altitude of 200 m; a further 30% lives between 200 and 500 m above sea level. There are many reasons why coastal locations have proved so attractive to large numbers of people:

- In common with the other hydrosphere stores described above, seas are both a source of food and a highly effective medium on which to transport large numbers of people and quantities of goods at a relatively small unit cost. The trend towards increasingly large ocean-going ships provides a classic example of the principle of **economy of scale**. Shipping trade creates a wide range of on-shore jobs, including the initial processing of imported raw materials and their subsequent manufacture into finished products.
- A high proportion of the world's most productive farmland lies within lowland areas adjacent to coastlines. The graph in Figure 4.25 shows that only 36% of the world's land area has genuine farming potential; much of this valuable land is located in coastal regions.
- Level or gently sloping lowland is ideal for the construction of buildings and transport networks. It is partly for this reason that the majority of the world's conurbations and megalopolises are located on or within a relatively short distance of a coastline.
- Coastal locations are favoured by retired and disabled people because many stretches of coastline are easy to walk on, have attractive scenery, and their on-shore breezes dilute and disperse polluted air. Sea currents have a moderating effect on coastal weather

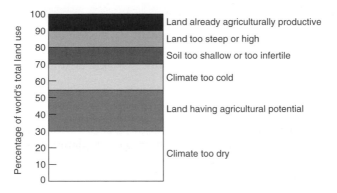

Figure 4.25 Suitability of global land area for agriculture

systems, and are therefore highly beneficial to people who cannot tolerate extremes of temperature.

- Many holiday resorts are in coastal locations, allowing visitors to enjoy traditional seaside pleasures as well as modern water sports.

Summary

- Every continent, country and region has a unique population distribution pattern due to a wide range of factors, including: latitude, climate, height and steepness of the land, soil fertility, accessibility and nearness to the sea, availability of water and other valuable natural resources.
- The most densely populated regions of Britain are Greater London and the coalfield areas; the most sparsely-populated areas are the mountainous regions of Wales and north-west Scotland.
- The most densely populated area in Brazil is the resource-rich coastal region in the south-east; the drought-prone north-east and the densely forested Amazon basin are sparsely populated.
- The development of individual settlements depends on the suitability of their site and situation.
- The settlement hierarchy ranks various types of settlement in order of size, from hamlet to megalopolis.
- Variations in population density also take place within urban areas (e.g. very few people in the CBD but far more in the adjacent inner residential zones). Urban redevelopment has led to the need to accommodate over-spill populations in suburban areas and new towns. Ethnic segregation may also lead to variations in urban population density and distribution.
- Urbanisation has led to the growth of megacities and a wide range of problems such as inadequate and insufficient accommodation and a low level of quality of life; this has been the result of rural–urban migration rather than an increasing rate of natural increase by the indigenous population.

Student Activities

Research projects

1. a) Use atlas maps and Internet sites to investigate the national population distribution pattern of any two of the following countries: Australia, China, Egypt, Japan and South Africa.

   b) For each of your chosen countries: provide a detailed description of the national population distribution pattern, and give detailed explanations for both the general national pattern and any significant regional variations from that pattern.

2. a) Investigate the population distribution characteristics of your own local area.

   b) Display your findings on an annotated base map that shows the chief physical features and settlement location patterns of this area.

Examination-style questions

1. Study the table below, which shows changing proportions of employees within the three employment sectors of Britain and Brazil, then answer the following questions based on it:

   a) Compare the current employment structure characteristics of both countries.

   b) Describe separately the changes in employment structure that have taken place in each country, then suggest reasons for these changes.

2. With reference to named settlements and localities within them, explain why population densities tend to vary considerably *within* large urban areas.

3. Population density patterns reflect the tendency for large numbers of people to cluster together. With reference to examples you have studied, describe the locational and economic reasons for the continuing increase in the rate of global urbanisation.

4. With reference to named examples, describe a range of strategies that have been used as part of urban renewal programmes.

5. Evaluate the validity of this statement: 'Physical factors are invariably more influential than human factors in the development of large urban settlements'.

	Brazil		UK	
Sector	2003	1971	2003	1971
Primary	10.10%	44.30%	0.90%	3.00%
Secondary	38.70%	18.40%	26.50%	34%
Tertiary	51.20%	37.300%	72.60%	63%

5 Advice to Students

Most readers of this book and its companion volumes in the *Access to Geography* series will be AS/A2 geography course students. The following additional introductory notes to the book provide advice that will help you to maximise the likelihood of achieving one of the higher examination grades. As in most academic disciplines, the best way of doing this lies in becoming proficient in each of the following:

- **Learning the subject-specific vocabulary** – the 'key terms' and their meanings. These terms are the building blocks on which the whole structure of the subject is based. Being familiar with them and knowing how to use them are both essential to the construction of quality examination answers. Quoting key words appropriately makes examiners aware of your familiarity with the subject and enables them to mark more positively. Each chapter of the book begins with some of the most crucial key words relevant to its topic; learn them as soon as you can, and become familiar with any other key words highlighted in bold print within each chapter.
- **Becoming familiar with case study material**. A case study is a set of facts about a particularly important place or event. For example, when asked to explain the suitability of Southampton's location as a major international port, it would be helpful to describe the city's sheltered location on Southampton Water and to the north of the Isle of Wight, its double-tides that provide deep-water access for an unusually long period of time, its large trading hinterland which includes the London conurbation, and its accessibility to both the continent of Europe and the Atlantic Ocean. Each of the book's chapters includes a selection of case studies and the sets of Student Activities begin with some recommended research topics that will enable you to add to your own collection of case study material.
- **Learning the subject material**. This involves revising both the subject-specific vocabulary and the case study material effectively. Each chapter provides a summary of its main contents.
- **Practicing answering examination questions**. Each chapter concludes with a number of questions of the type frequently set by AS/A2 Chief Examiners. Answering these questions as fully as possible – either under 'examination conditions' or with access to resource materials – is one of the most effective and diagnostic ways of raising achievement levels and increasing personal confidence in the ability to achieve success.

It is beyond the scope of any book to provide detailed case studies of every aspect of an advanced course of study and many relevant events are likely to take place after this book has been written. Students are, therefore, strongly advised to create their own additional resource banks. Doing this could be as simple as putting newspaper cuttings into four separate folders, each being devoted to the theme of one chapter of the book. This filing system can then be expanded later on by adding individual topic folders, e.g. a series of folders on mortality might contain separate information about HIV/AIDS, conflict-induced deaths and ethnic cleansing. When compiling such resources:

- Highlight the most important extracts *before* filing the cuttings. Highlighting at that stage forces you to read the articles and avoids mammoth reading-and-highlighting sessions much later in the course, when the top priority will be pre-exam revision.
- Be aware that many articles are written from a biased point of view and this is often done so subtly that it is often difficult to detect the author's 'hidden agenda'. If you do suspect major bias in an article, highlight just the facts because these are likely to have been carefully researched and so be more reliable than the interpretations and conclusions based on them!

Bibliography

All of the following recommended books are published in the UK, except where stated otherwise:

Ballard, R. (ed.) *Desh Pardesh: The South Asian Presence in Britain* (1994) Hurst.

Cohen, R. *Global Disporas: An Introduction* (1997) UCL Press.

Frampton, S. *et al. Natural Hazards: Causes, Consequences and Management* (1996) Hodder & Stoughton.

Gardiner, V. and Matthews, H. (eds) *The Changing Geography of the United Kingdom* (2000) Routledge.

Gillett, J. and M. *Physical Environments: A Case-study Approach to AS and A2 Geography* (2003) Hodder & Stoughton.

Guibernau, M and Rex, T. (ed.) *The Ethnicity Reader* (1997) Polity.

Guinness, P. and Nagle, G. *Advanced Geography: Concepts and Cases* (1999) Hodder & Stoughton.

Guinness, P. *Migration* (2002) Hodder & Stoughton.

Hardill, I. *et al. Human Geography of the UK: An Introduction* (2001) Routledge.

Hilderink, H. *World Population in Transition: An Integrated Regional Modelling Framework* (2000) Thela-Thesis, The Netherlands.

HMSO, *National Statistics – UK 2002: The Official Yearbook of Great Britain and Northern Ireland* (published annually).

HMSO, *National Statistics: Regional Trends* (published annually).

Jackson, S. *Britain's Population: Demographic Issues in Contemporary Society* (1998) Routledge

Jones, H. *Population Geography* (1990) Paul Chapman Publishing.

Lloyd, J. *Heath and Welfare* (2002) Hodder & Stoughton.

Mason, D. *Race and Ethnicity in Modern Britain* (2000) Oxford University Press.

Preston, S. *et al. Demography: Measuring and Modelling Population Processes* (2001) Blackwell.

Sarre, P. and Blunden, J. (eds) *An Overcrowded World?* (1995) Oxford University Press/The Open University.

Redfern, D. *Human Geography: Change in the United Kingdom in the Last 30 Years* (2001) Hodder & Stoughton.

Skinner, M. *et al. A–Z Geography Coursework Handbook* (2003) Hodder & Stoughton.

Skinner, M. *et al. Complete A–Z Geography Handbook* (2003) Hodder & Stoughton.

Tyler Miller Jr, G. *Living in the Environment* (1998) Wadsworth Publishing, USA.

Wills, C. *Plagues* (1996) HarperCollins.

Winder, R. *Bloody Foreigners: The Story of Immigration to Britain* (2004) Little, Brown.

Recommended Websites

The following selection of websites provides a starting point for personal research using the Internet. Each entry provides the 'home page' website, the name of the site provider and an appraisal of the information provided by that site relevant to the population-related topics covered by this book. Further site lists may, of course, be obtained by searching very general theme topics such as 'population' and 'migration', then refining them by progressively narrowing the search field in the usual way.

geography.about.com – *About*
Provides informative but readable articles on a wide range of topics. An ideal student resource, offering a range of services from blank world/regional maps to very detailed case study articles on both human and physical geographical topics.

www.bbc.co.uk – *British Broadcasting Corporation*
Useful source of topical news articles. Operates an alphabetical index search system. Best results are obtained by entering *very specific* details such as the precise dates and locations of events.

www.cdc.gov – *Centers for Disease Control and Prevention*
Provides a wide range of information about health matters such as cancer, transmittable diseases, immunisation, travel hazards and precautions, physical disabilities and the hazards of smoking.

www.census.gov/ipc – *US Census Bureau*
Provides global, regional and national demographic data such as population statistics, population pyramids, infant mortality, fertility, literacy and family planning programmes.

www.checkmyfile.com – *Checkmyfile.com*
Uses information derived from the UK's 10-year censuses to provide locality characteristics. Intended primarily for potential house-buyers, it summarises housing types, households' car ownership, age distribution of residents and the ethnic composition of populations.

www.cia.gov/cia/publications/factbook/ – *Central Intelligence Agency (USA)*
Particularly useful is 'Guides to Country Profiles', which provides a detailed summary of every country's key features and statistics, e.g. area, climate characteristics, natural resources, natural hazards, as well as population-related topics such as size, structure, growth rate, migration, life expectancy and HIV/AIDS status.

www.citypopulation.de – *City Population*
A specialised site providing mapped and tabled information about world, regional, national and city populations, and population trends.

www.dec.org.uk – *Disasters Emergency Committee*
Provides information about its work co-ordinating the activities of charities such as the British Red Cross, CAFOD, Christian Aid and Oxfam.

www.fao.org – *Food and Agriculture Organisation of the United Nations*
Provides global statistics on population, agricultural production and food balance sheets.

www.oneworld.net – *Oneworld*
A topical resource, providing details of news, campaigns and organisations involved in global human rights issues.

www.prb.org – *Population Reference Bureau*
A rich source of population-related data (at global-to-national scales) on family planning, fertility, poverty, migration, mortality and urbanisation topics.

www.refdesk.com/factpop.html – *Population and Demographic Resources – refdesk.com*
A worthwhile resource on a wide range of topics. 'Clocks' provide current global and national population estimates.

www.statistics.gov.uk – *National Statistics (an agency of the UK government)*
Migration and population are the key categories within this very wide-ranging data source.

www.unaids.org – *Joint United Nations Programme on HIV/AIDS*
Provides detailed information global and regional information about HIV/AIDS.

www.un.org – *The United Nations*
Very reliable source of information on current global and regional issues.

www.unesco.org – *United Nations Educational, Scientific and Cultural Organisation*
Provides global, regional and national data on cultural, education and literacy matters

www.who.int – *World Health Organization*
A reliable site that provides a wealth of detailed information about population, health and other global/national topics. One of its most valuable services is providing precise, up-to-date statistics about individual countries.

www.worldbank.org – *The World Bank*
Provides rankings of population, GDP and world development data. Also provides details of the HIV/AIDS states and information about its investment in HIV/AIDS programmes.

Index